IMAGES
of America

THE NAVY
IN NORCO

COWBOY SAILOR, C. 1945. Like many Norco Naval Hospital and Laboratory patients, staff, and scientists, corpsman Bob Allen stayed in the Corona-Norco area to build a life. Norco very likely exists because of the financial boon provided by the US Navy for over 60 years. It is fitting to see a Sailor atop a horse in Norco, a city known nationally today as Horse Town USA. (Jim Allen.)

ON THE COVER: THE BUCK STOPS HERE, 1952. Los Angeles newspaper the *Examiner* staged theatrical extravaganzas featuring entertainers, sports stars, and politicians "for the boys." Proceeds went into the newspaper's War Wounded Fund, which in turn passed cash to US Sailors and Marines wounded in the Korean War. Patients in front of the recreation center at the Corona Naval Hospital are holding sizable checks, which were clearly appreciated. (USC.)

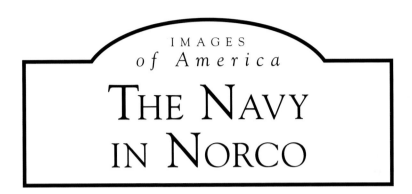

IMAGES
of America

THE NAVY
IN NORCO

Kevin Bash and Brigitte Jouxtel

ARCADIA
PUBLISHING

Copyright © 2011 by Kevin Bash and Brigitte Jouxtel
ISBN 978-0-7385-7526-1

Published by Arcadia Publishing
Charleston, South Carolina

Printed in the United States of America

Library of Congress Control Number: 2010941971

For all general information, please contact Arcadia Publishing:
Telephone 843-853-2070
Fax 843-853-0044
E-mail sales@arcadiapublishing.com
For customer service and orders:
Toll-Free 1-888-313-2665

Visit us on the Internet at www.arcadiapublishing.com

*To Rex Clark, whose magnificent Norconian—rather than
"folly"—was a haven of hope for thousands of wounded
and sick American leathernecks and Sailors*

ACKNOWLEDGMENTS

Thank you to the following libraries and archives that graciously allowed usage of their images: the Corona Public Library Board of Trustees (Corona), who also provided access to the *Daily Independent* archives; the University of Southern California (USC); the Herald Examiner Collection/Los Angeles Public Library (LAPL); the Sea Bee Museum (SBM); the Naval Surface Warfare Assessment Center Seal Beach, Detachment Corona (NSWACSBDC), who also provided the invaluable *History of the Naval Weapons Station Seal Beach*; the Bureau of Medicine and Surgery, Naval Surface Warfare Center (NSWC); Office of Medical History and Archives (BUMED); the Orange County Archives (OCA); Bison Archives; the Norco Historical Society (NHS); Academy of Motion Picture Arts and Sciences (AMPAS); and the Lake Norconian Club Foundation (LNCF). All photographs appear courtesy of the LNCF unless noted otherwise. When a photograph appears courtesy of a different institution, abbreviated credit is given. When a photograph appears courtesy of an individual, that person's first and last names are given. Unless otherwise noted, all quoted material comes from the *Daily Independent*.

Thank you to the late Ellen Revelle and the marvelous Peter Clark, who provided hundreds of his father's photographs, history, and great insight into his family's founding of Norco and building the Norconian.

Thank you to the following people for their stories, photographs, and assistance: the fabulous Jeanne Easum, Andre B. Sobocinski, Steve Mikesell, Huell Howser, Lorin P. Meissner, Orrin Anderson, Jerry Stewart, Jack Henderson, Elizabeth Kinzer O'Farrell, Eddie Musquez, Charles Roth, Don Stowe, Edna Johnson, Florence Rohleder Christensen, Florence and Ollie Marlow, Gina L. Nichols, Dottie Laird, Bill Erickson, Ralph Easum, Guillermina Hall, Michael Brownell, Dennis Ellison, Brian Davis, Nathan Lanni, Dennis Kent, Harrison Hueblein, William Lauper, Jon Jerges, Gregg Smith, Jay Kadowaki, Jennifer Marlatt, Trixy Betsworth, Norco historian Ron Snow, Dace Taub, Christina Rice, Mark Wanamaker, Betty Bash, Mary Winn, Velma Hickey, Bill Wilkman, Beth Groves, Jeff Allred, Kathy Azevedo, Jack Fry, the late Hal Clark, Ray Harris, Robert Peister, Gene Peister, Bob Allen, Jim Allen, Karlene Allen, Phil Newhouse, Lyle Draves, Jack Britton, Vickie Draves, Regina Dotson, Jo Ann Gordon, Debbie Stuckert, Sammy Lee, Rudy Ramos, Benjamin Cricket, Mike Nugent, Rueben Lemus, Lolita Lemus, Augie Ramirez, Larry Key, Pat Barker, John Barr, B.J. Hill, George Milner, Eunice Richardson, Shirley Gilbriath, Harold Davis, Jack Gordon, Don Williamson, Howard Hansclik, Red Taylor, Lesley Mathews, and Frank and Edna Day.

INTRODUCTION

In 1941, 60 miles from the nearest ocean, the United States Navy docked in Norco, California. The small agricultural community and the neighboring lemon hub, Corona, have never been the same. Uncle Sam was just entering World War II and hospitals for incoming wounded were in short supply. Extravagant health resorts built during the roaring 1920s were plentiful and deemed perfect settings for medical facilities. Resort owners across the nation, crushed by the Great Depression, were eager to sell for pennies on the dollar. Rex Clark, the founder of Norco and the builder of one of the most breathtaking resorts in America, was no exception.

Rex Clark was a dreamer with access to an unlimited pot of money. He dared, against all logic, to plant a blindingly magnificent recreational resort known as the Norconian Resort Supreme in the middle of a sea of chicken farms. Incredible success followed the 1929 grand opening, as the wealthy and brightest movie stars of the day flocked to Clark's so-called shining palace on the hill.

Success was short-lived. By mid-1940, after a decade of tax problems, mechanics' liens, and too few customers, the resort sat empty, nicknamed "Rex's Folly." The US Navy, however, saw the 700-acre complex as a perfect spot to establish United States Naval Hospital Number One.

In the wake of Pearl Harbor, famed architect Claud Beelman was entrusted with quickly converting the stunning Norconian into a state-of-the-art medical facility. The lavish hotel suites were converted into plush hospital rooms, the ornate ballroom became a temporary surgery ward, and Pearl Harbor survivors were served meals on exquisite china in the grand dining hall. The hot mineral baths were perfect treatment centers for those afflicted with polio and rheumatic fever, and the outdoor swimming pools, once reserved for Olympic champions, movie stars, and the richest of the rich, became a haven for wounded and shell-shocked leathernecks and Sailors. The chauffeurs' quarters were ideal housing for hardworking corpsmen, the lakeside Pavilion a fabulous Officer's Club, and Lake Norconian a refuge for battle-weary combat veterans just wanting to toss a line in the water and fish.

As casualties mounted, more beds were needed, and a massive building program commenced. A five-story hospital, masterfully connected to the old Norconian Hotel, was constructed. First Lady Eleanor Roosevelt dedicated the new addition that proudly flew the Navy E Award flag, signifying extraordinary excellence in wartime construction and purpose.

On the southeast side of the lake, a state-of-the-art, self-contained tuberculosis hospital was built. In these beautiful wards, thousands were quarantined and treated with groundbreaking techniques and medicines. On the northwest corner of the property, another self-contained hospital was built to treat those afflicted with rheumatic fever, malaria, polio, and venereal disease. Over 5,000 patients were treated in this facility, and again, medical advances were groundbreaking.

As thousands of patients poured in, additional staff was needed. In record time, Spanish Mediterranean–style housing for nurses, corpsmen, and the newly established WAVES was built—all artfully blending with the historical Norconian buildings.

Entertainment and recreation were deemed crucial to recovery and a marvelous gymnasium and theater complex were built. The gymnasium became a sanctuary where leathernecks and Sailors with useless legs found out they were still men through the game of wheelchair basketball. In the theater, patients were treated to first-run movies at no charge. Jack Benny, Spike Jones, and Bob Hope broadcast sidesplitting radio programs.

Another sanctuary was built in the form of the multi-denominational St. Luke's Chapel, which was adorned with poignant, commemorative stained-glass windows donated by Americans from every walk of life.

John Steinbeck wrote, "Probably the most difficult, the most tearing thing of all, is to be funny in a hospital. Everything that can be done is done, but medicine cannot get at the lonesomeness and the weakness of men who have been strong," and also, "It hurts many men to laugh, hurts the knitting bones, strains at sutured incisions, and yet the laughter is a great medicine." Hollywood flocked to Norco to deliver exactly that medicine. Kay Francis, Claudette Colbert, Lucille Ball, James Cagney, and many other stars regularly visited and entertained patients. One former corpsman remembers well Bob Hope demanding that the officers go to the back of the theater and bring the "real heroes" to the front. When the greatest Navy in the world threw open the hospital doors to civilians afflicted with polio, cowboy stars Hopalong Cassidy, Roy Rogers, and even Trigger entertained, one by one, stricken children entombed in iron lungs.

Beginning in 1951, a new era began in the long hospital wards. Where world-class doctors once struggled to beat tuberculosis, a deadly Cold War was now being fought. Crucial guided missile research and development was being undertaken by thousands of the smartest men and women in the United States. In the process, they left behind a legacy of patents and innovations that in a very real way defended a nation and changed the course of the world. In 1964, a very select group of scientists were singled out and entrusted with the burden of independently assessing why Navy guided missiles were failing. For three decades, these individuals acted on the trouble and failure reports of the revolutionary Polaris submarine missile program, arguably the nation's single most important strategic weapon of the Cold War.

The Navy's arrival had a profound effect on Corona and Norco. The economic engine generated by the naval hospital likely saved Depression-savaged Norco, and both communities experienced dramatic growth as thousands of patients, medical staff, visitors, and construction workers flooded into the once-quiet agriculture centers.

After the war, both communities experienced a slow move away from their agricultural roots as many former patients, armed with the GI Bill and a new opportunity to get a higher education, stayed in the area. Even more dramatic was the arrival of hundreds of well-paid scientists beginning in 1951.

Before the Navy (the first US armed service to integrate) came to Corona and Norco, both towns had an open Ku Klux Klan presence. Hispanic laborers were segregated to their own school, denied housing in many areas, and only allowed to swim in the local plunge one day a week, after which it was promptly drained and scrubbed down. Those barriers were toppled as all of Corona mourned their first Hispanic World War II casualty, and Navy Captain L.B. Marshall demanded that his African American patients, survivors of Pearl Harbor, be served in Corona restaurants.

Today, Norco and Corona are thriving communities, thanks in large part to the naval hospital that was disestablished in 1957. Incredibly, dozens of old hospital buildings are just as they were when built and likely represent the most complete example of a World War II–era naval hospital in existence. On the southern end of these historic grounds is the beautiful Norco College, and occupying the still beautiful former Norconian and naval hospital buildings is the California Rehabilitation Center; the Naval Surface Warfare Assessment Center Seal Beach, Detachment Corona; and the Naval Surface Warfare Center, Corona—70 years later, the Navy is still in Norco.

One

It All Started With Chickens

Norco, c. 1925. In 1920, Rex B. Clark purchased—or perhaps received as payment for a bad debt—5,000 acres just north of Corona, California. He had the idea to develop a community of self-sufficient chicken farmers who could produce enough food and income from their own land to feed a family. This was an odd dream for a man married into wealth and accustomed to living well. (Peter Clark.)

REX BRAINERD CLARK, 1925.
The founder of Norco and builder of the Norconian has been remembered as many things: a visionary, a buffoon, a racist, a philanthropist, and a scoundrel. He managed to finagle away a large part of his wife's trust fund in 1920 and spent unknown millions of her money on the resort family members referred to as Rex's Folly. The same place was eventually called paradise by thousands of combat-wounded veterans. (Peter Clark.)

BANKROLLED NORCO, C. 1925.
Grace Clark (left), daughter of publishing giant James Scripps, visits Norco, the township she likely financed, with members of her family. She might not be smiling if she knew her husband had more on his mind than chickens and was about to spend an even larger chunk of her inheritance to build the Norconian. (Peter Clark.)

NORTH CORONA LAND COMPANY, 1924. Rex Clark owned 51 percent of the construction, development, and sales component of his North Corona Land Company (NCLC) operation, with the main office in downtown Los Angeles. Through this company, parcels sold for around $800 per acre and $1,500 for five-acre plots. Here, an NCLC tractor-trailer rig (also sold regionally through the company) participates in Corona's Fourth of July parade. (Peter Clark.)

REX SCRIPPS CLARK, C. 1923. As director of the North Corona Land Company, Rex B. Clark's oldest son was instrumental in the early development of Norco and actually coined the town's name. Unfortunately, "Scripps" and his father did not see eye to eye on several fronts, and Scripps's wife preferred Detroit to rural Norco. So he moved on, but not before taking thousands of photos of Norco. (Peter Clark.)

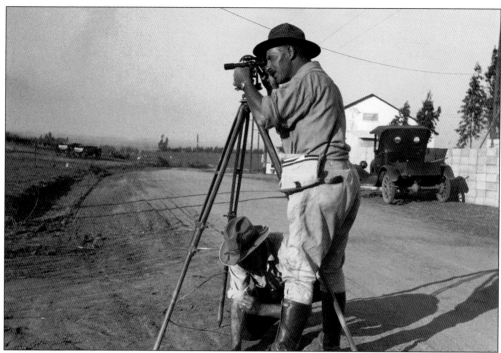

CUTHBERT GULLY, C. 1923. Norco's chief engineer for over 40 years was known as Captain because of his World War I service in the Army Corps of Engineers. As Rex Clark's right hand, "Cap" designed and built early Norco streets, reservoirs, pumps, and water delivery systems. A staunch patriot, Gully promoted "joining the fight" during World War I and was a founding member and long time commander of the Corona American Legion Post. (Peter Clark.)

DOWNTOWN NORCO, 1930. Norco's Orange Heights Mutual Water Company was located in the building to the right until the late 1950s. Upham's Drug store (left), a quick casualty to the Great Depression, was the first commercial building in Norco constructed as a non–North Corona Land Company business. After Norco's incorporation in 1964, this building became the first city hall and is today the town library. (Peter Clark.)

INCORPORATED UNDER THE LAWS OF THE STATE OF CALIFORNIA
DECEMBER 18, 1909   CORPORATION NO. 59428

Number 2599

---2.16--- Shares

Orange Heights Water Company

NORCO, RIVERSIDE COUNTY, CALIFORNIA

55,000 SHARES

CAPITAL STOCK $55,000.00

SHARES $1.00 EACH

This Certifies that Virgil E. Stambaugh, Sr., & Florence C. Stambaugh, husband and wife as joint tenants.

are the owner s of, and entitled to ---Two and sixteen/100ths---------------------------Shares

of the Capital Stock of the ORANGE HEIGHTS WATER COMPANY, which are appurtenant to the hereinafter described tract of land situated in the County of Riverside, to-wit: Lot Eight (8) Block Fifteen (15) of Norco Townsite, as per Map of said Tract, recorded in Vol. 11, page 4 of Maps, Records of Riverside County, California, containing .216 acres, more or less.---------------------

Said stock is subject to the By-Laws of said Company as the same now exist or may be changed from time to time, and such further rules and regulations as the Board of Directors of said Company may from time to time prescribe, and is transferable only on the books of the Company by endorsement thereon and surrender of this Certificate, together with a transfer of the land hereinabove described.

IN WITNESS WHEREOF, the said corporation has caused this certificate to be signed by its duly authorized officers and to be sealed with the seal of the corporation.

Dated this Twenty First day of June A.D. 19 45.

SECRETARY

PRESIDENT

ORANGE HEIGHTS MUTUAL WATER COMPANY, 1945. Until the late 1950s when a community services district was formed, every Norco landowner received 10 water shares per acre. Rex Clark, who retained all water rights within the Norco city limits, hoped for a big payday from water sales to the Navy hospital. Needing a more reliable source of water for its burgeoning medical facility, the Navy had other ideas. (Florence Marlow.)

GROWN IN NORCO, 1922. Prior to 1919, Norco was known as Citrus Belt or Orchard Heights, and was essentially a failed agricultural area. A faulty water supply, a location far from centers of commerce, and devastating Santa Ana winds created an inhospitable environment for all but the hardiest farmers. That changed when Rex Clark drilled wells, built reservoirs, and improved roads. By 1923, Norco was a poultry and agriculture Mecca. (Peter Clark.)

WALTER KNOTT, C. 1925. Norco became a leader not only in poultry production, but also in rabbits, lettuce, and a variety of other crops. Attracted by good soil and promises of plentiful water, Knott purchased land in 1924 to grow berries and sold them from this stand in Norco. The cost of water became prohibitive, so Walter consolidated his operation in Buena Park and built the famed Knott's Berry Farm. (OCA.)

THE NORCO STORE, 1923. The centerpiece of Norco's commercial zone opened for business on the township's dedication day. On the bottom floor, customers could find all manner of foodstuffs and dry goods, candy, ice cream, a butcher counter, and pharmacy. The Norco Grill and several apartments were on the second floor. Rex Clark, though proud of this store, lamented that Norconians purchased food in his establishment rather than raising it. (John Barr.)

PINT-SIZED PATRIOTS, C. 1938. The active Norco American Legion post sponsored multiple youth groups, one of which was the Girls Junior Auxiliary. Former member Edna Johnson Velthoen (far right) said, "We would go to meetings and they would teach us about patriotism and history. Later, when the hospital opened, we went to the dances to cheer up the fellas. Didn't hurt that they were cute." (Edna Johnson Velthoen.)

TOUGH TIMES, 1942. The Depression was devastating for the small town of Norco, and by 1940, many of these simple ranch homes—once so popular in the 1920s—were abandoned or terribly run down. At the time of this photograph, the poultry industry vital to Norco had collapsed and agricultural production had dwindled to dry farming. But the US Navy was on the way. (SBM.)

Downtown, 1941. By 1940, the Norco Grill was long gone, replaced by the Norco Library and several rooms for rent. The store below was still active, but with a lot less to sell. The Depression had reduced Norco's business center to the store, water company, gas station, and mechanic's shop. The drugstore, lumberyard, tile works, pipe company, and the rest were gone. (SBM.)

Rex and Emma, c. 1925. Shortly after his divorce from Grace Clark on October 8, 1930, Rex married his longtime mistress, Emma J. Snyder (right). Known affectionately as "Jimmie," Snyder was the longtime business manager of the North Corona Land Company and the unsung brains behind the development of Norco, the Norconian, and the deal made with the Navy to sell the resort. (Peter Clark.)

Two

Lemons and the Great Race

Corona, California, c. 1912. In 1893, Corona shipped its first oranges, and by the 1920s, this well-planned city claimed the title of Lemon Capital of the World. Its immaculate citrus groves stretched for thousands of acres, and lemons were picked, packed, and shipped worldwide. Before World War II, most residents were connected to the citrus industry, and the town weathered the Depression because the demand for citrus products remained strong. (Corona.)

THE CIRCLE CITY, 1919. H. Clay Kellogg created a one-mile-diameter circle with home parcels within and without. According to historian Mary Winn, "The founding fathers envisioned the circle being used for horse racing to make the city famous." As it turned out, a different kind of horsepower put Corona on the map: automobile races. During World War II, it was feared this landmark would be the perfect target for a Japanese attack. (Corona.)

CORONA ROAD RACE, 1913. Earl Cooper (pictured) was neck-and-neck with famed Barney Oldfield when a young boy stepped onto the track. Oldfield swerved, buckled a wheel, and broke his leg. Cooper stopped to help, but Barney called out, "Go on, you have a race to win!" Cooper won, with a world record, before thousands of spectators. Danger eventually ended the races, but not before putting Corona on the map.

18

MAIN STREET, 1936. Corona quickly established a central area of commerce that drew not only local residents, but regional customers as well. During World War II, this commercial zone received a huge economic boost as recuperating Sailors and Marines from the nearby naval hospital packed the street, requiring a very visible Navy Shore Patrol. Looking north, Norco's Beacon Hill, six miles away, is visible. (Don Williamson.)

MEXICAN-AMERICANS, 1924. Mexican workers were brought to Corona and Norco to tend citrus groves, dig ditches, and build roads. They were segregated to Washington School, small barrios, and allowed only one day a week in the community pool (after which it was drained and cleaned). With the first death of one of their own during World War II, the Hispanic community rebelled, and their lot was changed forever. (Peter Clark.)

Yippee Ti Yi Yo Little Doggies, c. 1936. The Corona and Norco American Legion Posts, each with large memberships of World War I vets, were extremely active between the two world wars, promoting patriotism through community events such as dances, sponsorship of baseball teams, and rodeos. Seen here is an American Legion pre-rodeo parade down Main Street. Both posts would contribute enormous amounts of time and money to the naval hospital. (Corona.)

Champions, 1939. Of the starting 11 for Corona High School, 10 would soon be fighting a world war, but in 1939, hated Colton High was the enemy. The players are, from left to right, (first row) Jim Ganahl, George Pauly, Austin Weit, Harry Higgins, Neal Snipes, Paul Snyder, and Clayton Wulff; (second row) Farrell Jones, Tony DeLeo, Martin Renfro, and Ed Hearn. All but Neal Snipes from Norco would come home. (Corona.)

Three

REX'S FOLLY

HOT WATER, 1925. Everything changed in Norco with the discovery of hot, bubbling mineral water. This photograph may well have captured Rex Clark (second from right) at the very moment he conceived the idea for "the greatest hot springs resort ever built." In any event, from this point on, chickens definitely took a back seat, and Clark was about to get into hot water in more ways than one. (Peter Clark.)

HUMBLE BEGINNINGS, 1922. Haystacks will soon give way to the manmade Lake Norconian, and the magnificent clubhouse will be built on the hill to the rear. This photograph perhaps illustrates one Navy land appraiser's observation as to why the resort failed: "Putting the Norconian in Norco was the equivalent to building the Woolworth Building in the Colorado Desert." For more information about the resort, readers can refer to *The Norconian Resort* by Kevin Bash and Brigitte Jouxtel (Arcadia Publishing, 2007). (Peter Clark.)

WHITE PALACE, 1929. This is the same view seven years later with the spectacular clubhouse to the rear and the Lake Pavilion to the right. Clark envisioned a "glistening white Palace rising from the sand." He also envisioned it as white in other ways: the term *club* was specifically used to discriminate against minorities and Jews. It was later speculated that this racist stance was yet another cause for failure. (SBM.)

PRIZED INVITATION, 1931. On February 2, 1929, thousands attended the grand opening of what was arguably the finest resort of its kind on the West Coast. The resort featured a manmade lake, airfield, first-class golf course, sumptuous dining room, lounge, ballroom, power plant, chauffeurs' quarters, laundry, and beautifully appointed rooms. The final tab for the resort—though difficult to confirm—likely ran as high as $4 million.

Rex B. Clark

Announces the Formal Opening of the
ᏞᎪᏦᎬ ᎤᏒᏣᎤᏁᏆᎪᏁ ᏣᏞᏌᏴ

THE MOST DELIGHTFUL AND UNUSUAL RESORT IN THE WEST
WHERE CLIMATE, SCENERY AND HOSPITALITY ARE THE BEST

on Saturday, February the second
nineteen hundred twenty-nine
Your presence will grace the occasion

Buffet Luncheon–Tea House – Twelve to Two o'clock
Sightseeing and Outdoor Amusements–Two to Six o'clock
Dinner–Eight o'clock — Dancing–Ten o'clock

Ten Dollars per Cover

THE CLUBHOUSE, 1929. Architect Dwight Gibbs designed this magnificent Spanish Mediterranean–style building, and famed designer A.B. Heinsbergen created the interiors. It was said to be earthquake-proof, and inside, guests enjoyed a luxurious hotel with dining facilities, a ballroom, lounge, and wonderful sweet-water spas. Handmade wrought-iron light fixtures, exquisite tile work, and stunning ceiling paintings are still breathtaking despite 85 years of use. (Corona.)

THE BIGGEST STARS, 1929. Hollywood flocked to the Norconian not only as guests, but also to film. Here, Norma Shearer and Robert Montgomery sit on the edge of the Norconian diving pool between takes for the tearjerker *Their Own Desire*. Will Rogers, Mae West, Randolph Scott, Robert Young, Harold Lloyd, Ann Sothern, and many more stars shot films at or around the resort. (AMPAS.)

BABY RUTH CANDY DROP, 1929. As a promotion, the famed candy company toured the country dropping Baby Ruth bars tied to parachutes from an airplane. In 1929, they arrived in Norco, and Rex Clark used the stunt to promote his Norconian airfield. He felt air travel was the future and envisioned planes arriving daily. Resident Polly Herbert poses in front of the Baby Ruth plane. (Robert Herbert.)

GOLF AT THE NORCONIAN, 1936. Well-known character actor Edward Arnold (right) poses in front of the Norconian Clubhouse; however, he was not actually on the original course, but a smaller, nine-hole course created in the 1903s. The once-immaculate links, designed by famed John Duncan Dunn, were an early victim of the Great Depression, and the course closed in 1933 with only sporadic use thereafter. Eventually, the Navy would reopen nine holes for recovering patients to enjoy.

LAST GASP, 1930. Balloon jumping over Lake Norconian was just one of the attempts made by Clark to attract attention and customers to his struggling resort (note the clubhouse in the background). In the midst of tax troubles, mechanics' liens, and empty bedrooms, the Norconian closed in 1933. An infusion of cash from his former wife's trust fund re-opened the doors in 1936 for one last hurrah.

GOODBYE, 1940. A name change from the Lake Norconian Club to the Norconian Hotel in an attempt to erase the stigma of racial exclusivity failed to save the resort, as did a final effort as Clark's Hot Springs with the resort touting the curative powers of the hot, sweet sulfur water. In truth, the once-immaculate resort had been for sale since 1933, and with no takers, the Norconian closed definitively in 1940.

SOLD! In 1941, eight days after Pearl Harbor, it was announced that the Norconian would be converted into a naval hospital. Secret negotiations had been going on for months and some speculate this deal was further proof that President Roosevelt had knowledge of the sneak attack before it happened. In any event, Norco and Corona would never be the same. Pictured here in 1951 is Florence Christenson, whose first job would be at the hospital. (Florence Rholeder Christenson.)

Four

THE NAVY DOCKS
IN NORCO

US NAVAL HOSPITAL, 1941. Rex Clark stated in the *Daily Independent* of December 15, 1941, "Extensive repairs have been underway at the hotel for some time in anticipation of the government taking over." Corona and Norco immediately went into support mode for the hospital and incoming patients by creating an active USO, a transportation network, burgeoning naval auxiliaries, and eventually, a large corps of Red Cross hospital volunteers known as Gray Ladies. (SBM.)

CLARK VS. US NAVY. On November 8, 1941, President Roosevelt set aside $2 million from his war emergency fund to purchase the Norconian. On November 11, 1941, Admiral Ross T. McIntire, surgeon general of the Navy, dickered unsuccessfully with Rex Clark at the White House to bring down the price. On December 3, 1941, the first Navy appraisal came in at $2.25 million and President Roosevelt subsequently completed the so-called "bargain" by executive order. Five months later, Clark still had not been paid and in July received notice of a condemnation suit to determine what the federal government would actually pay for the Norconian. These two southerly views show the pre-sale (above, taken around 1929) and post-sale Norconian (below, 1945). Significant construction had occurred throughout litigation with the addition of the Unit I Hospital (center), nurses' quarters (center right), and the Recreation Center (middle right). (SBM.)

CLARK SINKS NAVY. A subsequent Navy appraisal devalued the property—now termed a white elephant—to $725,000. Clark's business skills were assailed, and the "overbuilt resort" set in Norco was compared to a hotel that "had been built in a malarial swamp." Clark in turn accused politicians of killing the deal because he had refused their demands for kickback payments. Clark prevailed in court and claimed receipt of the entire $2 million; however, other reports suggest he received as little as $850,000. Seen above is a mid-1941 appraiser's aerial photograph of the Norconian before it was sold. Below is an image from 1944 that captures the dramatic expansion of the hospital, which by then included specialized units for the treatment of catastrophic wounds, malaria, rheumatic fever, tuberculosis, tropical ulcers (referred to as jungle rot), venereal disease, and general medicine. (SBM.)

ONE HOSPITAL, THREE NAMES. In December 1941, the facility at the Norconian was designated Naval Hospital Number One. On January 6, 1942, the name became the US Naval Hospital, Norco. On February 2, 1942, because Norco was not incorporated, the nearest train depot was in Corona, and the base postal service was technically through Corona, the facility was re-designated the US Naval Hospital, Corona. The change confused many and irritated Norco residents for the next several decades.

RESORT TO HOSPITAL, 1942. In February, famed architect Claud Beelman was retained "for the conversion of the Norconian Hotel to a hospital." The changes were extensive, and additions included four large operating rooms on the third floor, some of the finest X-ray equipment in the country, laboratory and dispensary facilities, wheelchair ramps, and a bed capacity of 268. Located on the first floor was the base store and Ships Service.

FIRST TO ARRIVE. On February 18, 1942, it was reported that the hospital had two patients. Two days later, 24 more arrived; most were survivors of the Pearl Harbor attack. The hotel was in the midst of conversion to a hospital (note the boards serving as wheelchair ramps), but despite the construction, one Sailor remarked, "Mom and dad will not believe where I am right now." (USC.)

NOTHING BUT THE BEST, 1942. The first patients were moved right into the resort bedrooms and treated by two teams of Mayo Clinic doctors, who volunteered for $1 per year. Each doctor was a specialist, and their expertise included plastic surgery, burns, tuberculosis, malaria, wound treatment, and others. Here, several Marines lounge in a suite that may well have once served a movie star or millionaire. (USC.)

BALLROOM TO LIBRARY, 1942. The fabulous Norconian ballroom (north view above, south view below) was the scene of many a party, and not a few movie stars danced deep into the night on the sprung floor. The Navy initially utilized this beautiful room as a ward for the first bedridden patients to arrive at the hospital. This Art Deco delight adorned with dozens of ceiling paintings also saw service as a makeshift surgery unit, theater, library, and—every once in a while—as a USO dance location. One woman from the Navy WAVES remembered well dancing with "a very short and freckled" Jimmy Cagney, and a patient later wrote of a "kiss for your mother" received from movie star Claudette Colbert. This room has not changed in almost 90 years; however, water is slowly destroying this one-of-a-kind masterpiece. (Above, SBM; below, BUMED.)

DINING ROOM, 1941. When an appraiser from the federal government took the above photograph in 1941, the Norconian Resort was closed for good. With the closed curtains, perfect rows of antique laced-leather-bottom mahogany chairs, and unlit wrought-iron chandeliers, this image exudes sadness. A few months later, the Navy filled this wonderful room with Marines and Sailors who chowed down on genuine china plates using the finest silverware money could buy (right). For over 80 years, this room has played host to movie stars, kings, queens, and dignitaries from all walks of life; heroes from the Bataan Death March, Tarawa, and Iwo Jima; rocket scientists and famed physicians; and finally, convicts. Today, despite being abandoned and neglected, the dining room is virtually unchanged—apart from the slow, unnecessary, and tragic destruction by water. (Above, SBM; right, USC.)

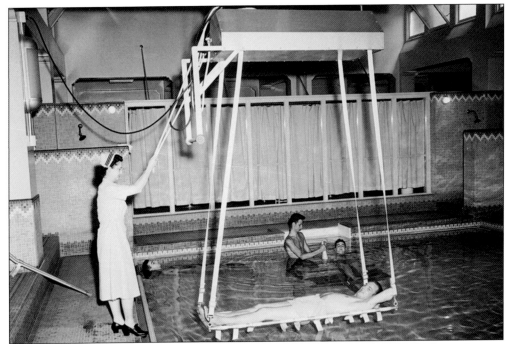

WHAT A LIFE, 1942. This indoor pool is surrounded by a complex of tiled cubicles and rooms where the rich and famous could receive a Turkish, Roman, or Russian bath; a mud, oil, or alcohol rub; a steam box treatment; and much more. The Navy used it as a polio and rheumatic fever rehabilitation facility. (USC.)

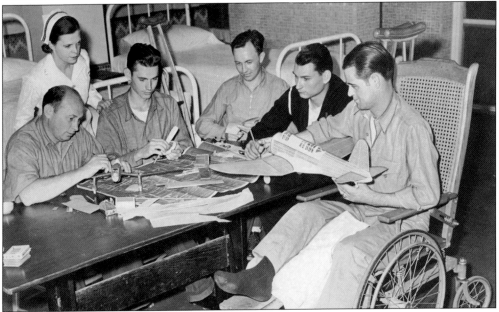

HOT SPRINGS BUNKS, 1942. These Pearl Harbor survivors engage in physical therapy to regain hand dexterity by assembling model airplanes. As the resort was being converted, patients actually slept in the spa areas, as shown. Decorated for his bravery at Pearl Harbor, Dale Lyons (far left) was a guest of honor at Corona's first community effort to sell war bonds and raise relief funds for the Navy. (USC.)

WORLD-CLASS WATERS, 1942. These patients dip their toes into what were once the only outdoor Amateur Athletic Union–qualifying competition diving and swimming pools in Southern California. Several international, national, and state swimming records were broken in these waters, and many Olympic gold medal heroes demonstrated their skill from these diving platforms. During World War II, famed divers Sammy Lee and Pat McCormick gave demonstrations for the patients. (USC.)

PEARL HARBOR HEROES, 1942. Sitting in the former hotel lobby, J.G. Raines (left) braved flames to rescue two crewmates during the Japanese attack. In an act of heroism, Dale Lyons (center) lost a foot during the onslaught, but eventually returned to duty with a prosthetic. Lyons later became the prime drone pilot in the Navy and operated an F-80 during the Bikini Atoll atom bomb tests. (USC.)

Captain H.L. Jensen, 1945. Famed senior medical officer of the USS *Solace*, Captain Jensen (right), was appointed the new hospital's first medical officer in charge and eventually oversaw what many considered to be the finest naval medical staff of the war. Jensen, in the hospital parade area, receives a Navy Unit Commendation for Meritorious Service for his heroic actions during the attack on Pearl Harbor. (LAPL.)

Nation's Best, 1942. By May, 100 Pearl Harbor survivors were recovering in Norco, most from the *Arizona*, *Utah*, and *Oklahoma*. Captain Jensen (ninth from left) remarked that when the battle started, 50 patients were onboard the USS *Solace*, and despite bringing aboard dozens of wounded, there were never more than 50 patients, as injured men ignored orders and rose from sick beds to assist rescue efforts. (USC.)

SHIP THAT NEVER SAILS. Pictured is the former Norconian Complex, consisting of a hotel, twin spas, pools, and a powerhouse. The *Los Angeles Times* wrote in 1942, "The patients, doctors, and nurses talk ship jargon. All 'hands' eat in the 'galley,' one goes 'ashore' or 'comes aboard,' not leaves or enters the hospital. The shifts are called 'watches' and so on."

EXPANSION, 1943. A sum of $1.5 million was allotted from Naval Appropriation Act funds for three new ward buildings (right center) connected to the original hotel (center), nurses' quarters (upper center), corpsmen's quarters (lower right), and water and sewer system improvements. Additional temporary wards are being laid out in the upper right; with their completion, 1,000 beds would be available. (SBM.)

ARMY-NAVY E AWARD, 1943. The quality of completed construction earned contractors Guy F. Atkinson and George Pollock the prestigious Army-Navy E Award for "excellence in war production." Atop the building, just beneath the American flag, the red and blue E Award flag flew, and each worker received a miniature E Award lapel pin. Executive officer Dr. Walt Waltman stated, "Your efforts here have been dedicated to the saving of lives."

ELEANOR ROOSEVELT, 1943. The president's wife dedicated the new hospital wing and remarked, "It was less like a hospital than other similar establishments she had visited," and that "it is indeed an admirable place for the Navy to take care of its men." The first lady (left, in hat) is having lunch in the former Norconian dining room with Captain Jensen (center) and his wife (right center). (Florence Marlow.)

Nurses' Quarters, 1945. Architect Claud Beelman was assigned the task of blending new buildings with the existing Norconian masterpiece, and met the challenge by matching the red-tile rooflines of the old resort and still maintaining his own unique style. He was able to align the two wings of the nurses' quarters with the existing Norconian Hotel to great effect.

Nurses, c. 1945. Prior to 1941, Navy nurses, such as the two pictured, were treated as officers socially and professionally, but not formally recognized as commissioned officers until World War II. Sadly, this photograph is one of hundreds that have been found in various albums depicting nurses, doctors, staff, and patients posing at the hospital with no identification whatsoever. These nurses' names and their stories could be lost forever.

First Flight, April 23, 1944. From the ACME news service: "In the first transcontinental flight of a naval hospital plane, 14 Sailors and Marines were flown from Washington airport to the Naval Hospital in Corona, California. The men are all suffering from rheumatic fever. Here Mrs. Harold Byrne, bids goodbye to her husband, seaman, first class Harold Byrne."

Wanted! Wanted!

CONSTRUCTION WORKERS
FOR THE U. S. NAVY

WEST COAST COMMAND CONSTRUCTION
LABORERS
60 Hour Week

Free Transportation and Expenses to Job
Representative Now Hiring

U. S. Employment Service
Address 3469 Main, Riverside, Calif.

Representative at U.S.E.S. Office, Corona City Hall
on Fridays

Wanted, 1944. The need for construction workers during World War II was acute, as most able-bodied men were off to war. With literally millions of Americans juggernauting forward to fight the Axis powers, facilities to house, feed, train, and care for soldiers needed to be built quickly; the naval hospital was no exception. Red Taylor, future mayor of Norco, worked the hospital construction project "from start to finish." (*Daily Independent*.)

CORPSMEN'S QUARTERS, 1945. This four-wing housing unit has had many uses over the years: during World War II as corpsmen housing, in 1959 as a missile systems assessment laboratory and storage for several historic computers, and in the 1970s as home to the George Ingalls Memorial Armory. Today, this very historic building is abandoned and deteriorating with the only activity being police training exercises.

NAVY CORPSMAN, 1942. Pictured is Jack Britton with the hospital mascot, nicknamed "Rags." Britton served at Norco from February through May in 1942 and was one of the first naval personnel to arrive at the facility. Britton remembers the grounds being beautiful and that his duties consisted of everything from assisting patients to raking leaves. Many corpsmen trained in Norco prior to serving in battle units overseas. (Jack Britton.)

HOT TIME, 1942. Initially, corpsmen stayed in the pictured old Norconian chauffeurs' quarters, which became the scene of some very significant parties. One Navy corpsman recalled waking up at the top of a tree with only his sailor cap on. Make no mistake, in combat, Navy corpsmen displayed unique courage and were revered by battle-tested Marines. (Jack Britton.)

OFFICERS CLUB, 1944. The Navy converted the former Norconian Pavilion into the so-called shipboard officers club. Some wild parties also took place in this wonderful lakeside building. As one story goes, a captain who shall remain nameless decided in the midst of partying to go fishing. Well, the boat sank, but the captain calmly awaited rescue in the shallow lake with his head, cigar, and bottle just above the water line. (SBM.)

QUITE A SPREAD, C. 1948. The pictured dining room is virtually unchanged from the day it served guests of the Norconian. Many Navy Christmas, Thanksgiving, Easter, and other special occasions were celebrated in this marvelous room; in fact, they were a tradition. While the actor Louis Hayward was a patient, WAVES member Vickie Dean recalled his wife, actress Ida Lupino, scrubbing these floors and decorating in preparation for a party.

LUSH LIFE C. 1946. To decorate his resort, Rex Clark purchased a Pasadena nursery and moved the entire stock of thousands of trees and plants—and even the former owner—to Norco. The nursery facility was located at the corner of Fifth and Hamner and today is known as Neal Snipes Park, named for a Norco boy killed at Tarawa. Here, a Sailor enjoys the former Norconian gardens.

43

TUBERCULOSIS UNIT, 1944. The Associated Press wrote, "Once one of the most beautiful golf courses in California, it is now one of the US Navy's great West Coast hospitals." The tuberculosis hospital complex, designed by Claud Beelman, was built on the east side of the former Norconian links. The first phase of construction consisted of a permanent facility with a capacity of 250 beds. Rather than construct a traditional sanitarium with everything under one roof, the building, known as Unit II, utilized long, separate hospital wards allowing fresh air to flow between. Each ward had glass-enclosed porches and was connected by a wide, covered walkway. This treatment complex was a single, integrated city and included its own cafeteria, morgue, commissary, theater, and staff quarters. Above is a northwest view of the newly completed Unit II, and below is an east view of one of the wards. (SBM.)

DEATH VALLEY, 1944. Duty in Unit II was feared because contact with infected patients was extremely dangerous, as tuberculosis was easily transmitted, and before antibiotics, it was a death sentence. Some suffering patients did in fact make the short walk to Lake Norconian and drowned themselves. Corpsman Ollie Marlow well remembers searching the murky waters for a suicide. Here, an unidentified nurse is taking every precaution to avoid infection.

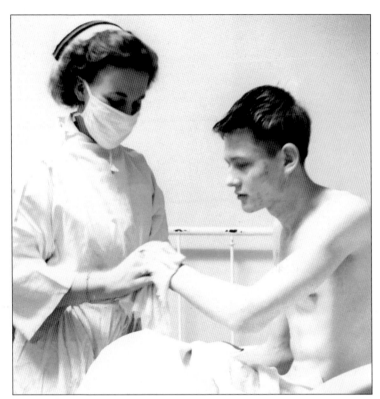

PIONEERING BREAKTHROUGHS. In October 1944, it was reported nationally that the naval hospital had successfully used penicillin to combat tubercular complications. In 1947, as the designated West Coast naval tubercular facility, Norco was selected to work with Olive View Sanitarium, one of three facilities in the nation to receive dosages of streptomycin for use in pioneering treatment of TB. Another unidentified nurse poses in front of the TB wards.

RECREATION CENTER, 1945. In 1943, the Navy announced a $4 million expansion to increase the number of available beds by 1,000. Also to be built were a recreation center and gymnasium, a theater, a chapel, WAVES quarters, a series of Moduloc wards, and an addition to the existing TB wards. Seen above under construction are the gymnasium (left) and theater. The proscenium opening, stage, and fly space for curtains are apparent by this time. In the view below, taken six months later, the gym (lower middle), the 1,200-seat theater (center), and the recreation rooms (lower right) were ready for recuperating Sailors and Marines. The state-of-the-art theater could accommodate large theatrical productions, movie showings, and radio broadcasts. The gym consisted of two full basketball courts, and the recreation center offered pool tables, a bowling alley, exercise equipment, and board games. (SBM.)

ROLLING DEVILS, 1946. While there are differing opinions as to where wheelchair basketball began, there is no doubt that teams organized in Norco were among the first, and that the Rolling Devils brought national attention to the sport and the men who played it. Johnny Winterholer, Bataan Death March survivor and star of the team, is fifth from right. (Elizabeth Kinzer O'Farrell.)

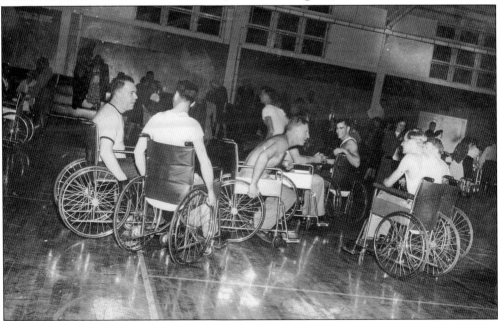

PIONEERS, 1946. Lt. Kinzer O'Farrell stated, "Treating the whole patient rather than just his disease or disabilities was mandatory and became the modus operandi in caring for the paralytic patients at Corona Naval Hospital." To help combat bedsores and infection among paraplegic patients and create healthy exercise, permission was granted to allow wheelchair basketball in the main gym. (Elizabeth Kinzer O'Farrell.)

HISTORIC, C. 1947. Don Stowe (left) supervised the gym and likely refereed the earliest wheelchair basketball games. According to Lt. Kinzer O'Farrell, the program generated so much interest that patients and staff began dropping by the gym to watch and applaud. As the first organized game approached, she stated, "Suddenly, we were seeing our patients beginning to feel like men, not like cripples." When asked about the first game, Don Stowe said, "The place was packed." (Don Stowe.)

THE SHOW MUST GO ON, 1947. Don Stowe, corpsman, third class, is shown transporting movies from the main theater to the tubercular wards. He served as a motion picture operator and gym supervisor, which he preferred over "bedpan commando." He operated base projectors on a nightly rotation with two other projectionists; other tasks included ambulance duty and the occasional special watch. (Don Stowe.)

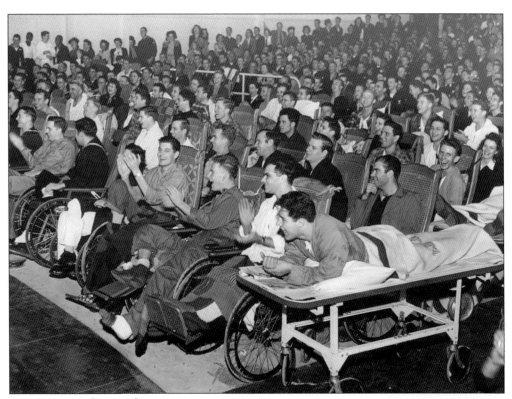

Al Malaikah Shrine Christmas Party, 1945. The Los Angeles–based Shriners were quite active at the Corona Naval Hospital throughout the facility's 17-year history, supplying radios, blankets, and gifts; they also sponsored a gala with a Christmas show for the opening of the new auditorium. Clearly, the boys had a great time. Pictured in the second row on the far left is Charles Roth, a former patient and terrific information source. (LAPL.)

Red Skelton, 1945. The redheaded MGM comedian was a regular at the hospital and is seen here with Edna Borzage as the two wait to go on for the "star-studded show presented by the Shriners in the new auditorium." According to the hospital newspaper, the *Corona Beacon*, Skelton "had everyone in the aisles with his jokes, fast patter, and inimitable imitations." (Academy of Motion Picture Arts and Sciences.)

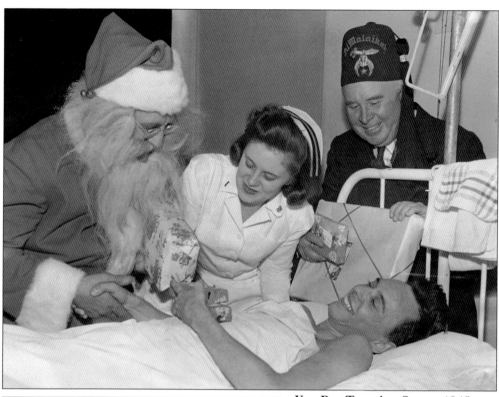

YOU BET THERE'S A SANTA, 1945. Marine gunnery sergeant Stance Polak from Wisconsin receives gifts from Santa while Charlotte Murphy and the Shrine's chief rabban, Vierling Kersey, look on. The Al Malaikah swept through the naval hospital the day after their Christmas show and handed out hundreds of gifts, a tradition carried out in Norco for almost 15 years. (LAPL.)

CORONA BEACON
GUIDING FOOTSTEPS TO A HAVEN

April 20, 1946   U.S. NAVAL HOSPITAL, CORONA, CALIF.   Vol. 1, No. 24

HELEN KELLER VISITS USNH, CORONA

• CITATIONS •

Among the heroes decorated by Captain F.C.Greaves (MC) USN was Lieutenant Commander Harry Albert Barnes (MC) USNR, who received the Bronze Star and a Citation in the Citation Ceremony held on Saturday morning, 13 April, 1946.

Lieutenant Commander Barnes received his award for meritorious service while on duty as Operating Surgeon in the Medical Department of an assault transport, during the assault on Okinawa. During this period, his department treated one thousand four hundred combat casualties.

Pharmacist James F. Bray received his second Letter of Commendation. This was given for outstanding services while interned as a prisoner of war at Bilibid Hospital for Military Prison Camps,Manila, Philippine Islands. Pharmacist Bray was a prisoner at Bilibid Prison from 6 May,1942 to 4 Feb'y, 19-45.

A Commendation and a Ribbon Bar were awarded Capt. Stanton T.Allison (MC) USNR for service as Director of Clinical Services on the U.S.S. Benevolence. That ship provided a screen- ing and hospital facility for the care of 1500 Allied prison- ers of war from the Tokyo Bay area.

Ensign Carl E. Fleming received his second Letter of Commendation and the Navy Unit Commendation for service in action as First Lieutenant of the U.S.S. Bennethead.

Corporal Harold S. Wilcox (USMC) re- ceived the Purple Heart for wounds re- ceived in the Asiatic-Pacific Area. As a result of wounds received during the Pearl Harbor attack Chief Carpenter Ray C. Smith (USN) also received the Purple Heart.

Chief Pharmacist Mate Jesse Royce Chazbliss was awarded the Presidential Citation for courageous conduct during the defense of Wake Island.

For extraordinary heroism during the battle off Samar,Phillippines, John Stephen Glova CM3/c received the Presidential Unit Citation.

Chief Pharmacist Mate Henry C. Trigg was awarded the Navy Unit Citation.

Miss Rosenthal, Pete Kirschner, Miss Flannery, Miss Keller, Miss Thomson.

We hear and read about some famous people, but never realize that there is a possibility of ever meeting them That was the feeling of the audience at USNH Corona when they met the world famous Helen Keller last week.

Miss Keller with her companion and in- terpreter, Miss Polly Thomson, lectured in the auditorium, then visited some of the wards. They sat and inspired all of the bed patients, especially our par- aplegics (spinal cord cases), for whom Miss Keller felt a great kinship.

Through the lecture of Miss Thomson, we heard about Helen Keller's life: the fact that she was born in Tuscumbia, A- labama, on 27 June, 1880, and that at the age of 19 months she contracted what is now believed to have been spinal men- ingitis, which left her deaf and blind, and the deafness resulting in her be- coming mute. Helen's world had been

shut out before she had began to live, but Miss Anne M. Sullivan (of the Per- kins Institute for the Blind, Mass.) was secured as a teacher for Helen, and opened her mind to all the knowledge that the world had to offer. Miss Sul- livan had only 1/10 normal vision all of her life, so she understood the greatness of Helen's handicaps.

Slowly the education of Helen Keller began to take form. It was with a doll that Miss Sullivan first taught her; that is, she would spell d-o-l-1 on her small hand, and would not relinquish it until Helen had spelled the word. Pre- vious to this, Helen's only method for expression was a nod of the head (for yes), or shakes of the head (for no), or imitative actions such as smacking her lips for candy, etc.

As she was learning, Helen could not seem to understand that everything had

(cont'd. on P.2)

INSPIRATION, 1946. Hospital newspaper the *Corona Beacon* covered the visit of Helen Keller in April. According to Keller's escort, Navy WAVE Vickie Dean, "Miss Keller moved from ward to ward touching the faces of patients and letting them touch hers. She was the most inspiring visitor I saw while stationed at Corona. The patients loved her." Later, Keller spoke through her assistant in the new auditorium. (Charles Roth.)

50

St. Luke's Chapel, 1946. The first service in this beautiful chapel was held on Christmas Eve 1944. Two naval chaplains, a rabbi and a priest, were in residence during World War II. After the war, local pastors were at times allowed to hold community services, and literally hundreds of weddings, christenings, and funerals took place under the arched beams of this house of worship.

Stained Glass, 1944. Adorning the interior of the chapel are 10 stunning stained-glass windows commemorating fallen heroes, corpsmen, nurses, and others. Still beautiful today, the windows were created and installed by famed stained-glass master Fred Wieland and donated by businesses and groups such as Barker Brothers Furniture, hospital staff, Navy mothers, and the ambulance corps. (Brigitte Jouxtel.)

NURSE BETTY, C. 1946. Elizabeth Kinzer O'Farrell LT (NC) USN (RET) is pictured in the front center before St. Luke's Chapel. O'Farrell tells of her experiences in her delightful memoir, *WW II A Navy Nurse Remembers* (CyPress Publications, 2007). Regarding her duty working with paralytic patients in Norco, she stated: "Both as intern and graduate physical therapist," that she "never worked harder" or "enjoyed [her] work more." Eventually, Nurse Betty became the supervisor in the physical therapy department. (Elizabeth Kinzer O'Farrell.)

GATEHOUSE, C. 1946. Pictured is the Fifth Street entrance to the hospital. Shown is the Beelman-designed gatehouse, which even today beautifully blends with the red-tiled Norconian clubhouse. During World War II, Marine guards would man this post. It is amazing to think who passed through these gates: Eleanor Roosevelt, Joe DiMaggio, Bob Hope, and many other celebrities.

WAVES, 1945. Despite the controversy regarding women serving in the military, the newly created Women Accepted for Volunteer Emergency Service (WAVES) became essential to the Navy—and the hospital in particular—performing all manner of duties, even fire-fighting training. WAVES were very popular at weekly dances for patients who had not received leave. The quarters pictured under construction housed over 200 young women at one point. (SBM.)

OFF TO HUNTER COLLEGE, 1944. WAVES members were trained at Hunter College in the Bronx in New York. The *Daily Times* reported that "one hundred and twenty WAVES recruits from twenty different Southern California communities," including Corona and Norco, would train there. A young lady from Riverside stands in the middle of this image, which was taken across from the Los Angeles Union Station. According to Vickie Dean, "Those WAVE days in Norco were some of the best in my life." (LAPL.)

SPLINTERVILLE CONSTRUCTION, 1943. Unit III was part of the 1,000-bed expansion project and included 30 semi-permanent Moduloc buildings made of a wood plastine laminate. The buildings very quickly splintered, hence the nickname "Splinterville." This standalone hospital was connected by a long covered walkway and laid out geometrically on the northwest corner of the hospital property. Designed to last 10 years, this complex is actually still in use today as housing for prisoners. (SBM.)

HEALING HAVEN, 1944. The Navy's national rheumatic fever facility treated over 5,000 patients during World War II. The medical officer in charge was George C. Griffith, who later claimed the use of penicillin to treat rheumatic fever patients was pioneered in Norco. Victims of polio were also treated in these wards, where treatment was in part based on the work of unorthodox polio fighter Elizabeth Kenny. (SBM.)

EDDIE MUSQUEZ, 1946. Musquez, in front of the former Norconian, was diagnosed with rheumatic fever and sent to Norco, where he was given heavy doses of penicillin twice daily. After six months of forced wheelchair confinement, he was unbearably bored and actually sought duty scrubbing the magnificent Norconian floors. The sundial next to Musquez was purportedly a congratulatory gift to the facility from President Roosevelt. (Eddie Musquez.)

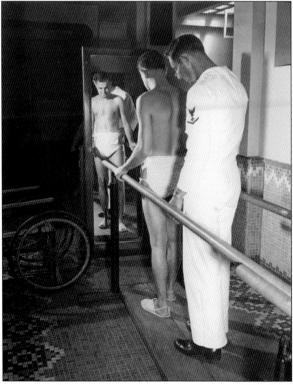

POLIO, 1946. Norco was the Navy's designated West Coast polio facility. Pictured is a patient undergoing treatment in the former Norconian spa. Famed polio fighter Sr. Elizabeth Kenny visited the hospital and presented her revolutionary treatment ideas. Norco polio patient Charles Roth stated, "While learning to walk one step or two at a time, we had on a really teeny bikini-like bottom that left us feeling naked." (USC.)

55

HONORED, 1946. Patients salute nurse Dorothy Behlow (center) with Kenny sticks, which were "crutches cut off about elbow-length with canvas straps to give a better grip." Behlow received a Unit III "Medal of Honor" for "meritorious service in the battle of Corona and the invention of the famous Behlow Bop." The referred-to battle was the fight to get well and the bop was the nurse's customary dance from patient to patient. (Charles Roth.)

SLOW DOWN, 1946. Charles Roth remembers that while wheelchair-bound, he "was used to racing down the hall and swinging into the elevator. One evening, I almost nailed Peter Lawford! He and Keenan Wynn had brought a piano and did a song-and-dance comedy skit for the few of us who were on the ward and hadn't pulled liberty." Roth is pictured with WAVES member Barbara Killen. (Charles Roth.)

PAIN, 1948. Charles Roth remembered, "Treatment started with hot packs made from cut-up blankets soaked in hot water and wrapped around our legs, arms and chest. This was very uncomfortable, but if we went a day without it, the pain let us know it was better to take the hot packs. They stood me upright 6 to 8 weeks after first being unable to walk and eventually taking one or two steps; we were learning to cope with what muscles could work and those that weren't going to work anymore. In the pool, we learned to walk in the water between two long poles (right). I always felt so lucky to not have my arms affected." By 1948, physiotherapy facilities included heat cabinets, whirlpools, specialized gymnasiums, and Hubbard tanks (below). (USC.)

57

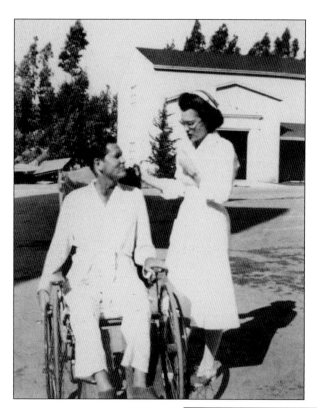

ONE TOUGH PALOOKA, C. 1946. In front of St. Luke's Chapel is Gene Allison with nurse Rose Mary Brueggeman. According to Charles Roth, "Gene had been on a cruiser at the start of the war when he got polio. He was sent to a hospital in Manila. When Manila fell, he spent the next three and a half years in a Japanese prison camp." (Charles Roth.)

PROSTHETICS, 1944. Thousands of World War II Marines and Sailors became amputees following shattering wounds. High visibility of these vets created a public outcry, prompting new awareness and major prosthetic advances. Norco-based doctors worked closely with several manufacturers to create better artificial limbs and pioneered physical therapy designed to get vets back to productive lives. (USC.)

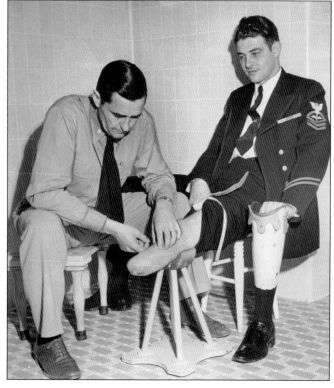

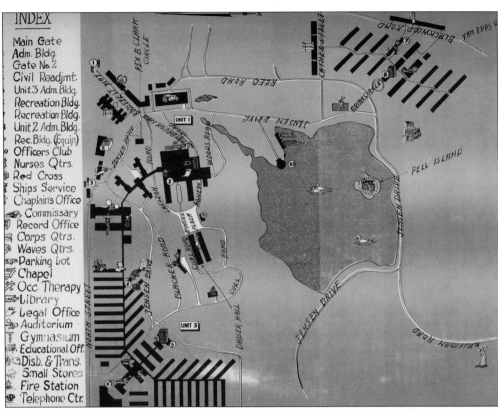

SIGNIFICANT NAMES, 1948. The hospital street names held national and local significance. Jensen Drive was named for Captain Jensen; Reeb Road for local hospital engineer Bill Reeb; and Corporal Pell Island was named for the pelicans that landed there and to poke fun at the Marines. Stambaugh Road was named for Commander Stambaugh, the only person who could back a truck up the steep road. (BUMED.)

GAIN FROM PAIN, 1944. The occupational therapy unit in Norco was crucial to developing strength and coordination in limbs and fingers debilitated by polio, rheumatic fever, and war-related injuries. The first tools received were from Richard Thew, who lost two sons in the Pacific. Thew initially sought revenge by enlisting, but was turned away as too old; so instead he donated tools to "rebuild men." (LAPL.)

59

THERAPY THROUGH ART, 1944. While a patient in Norco, famed artist Stanley Landsman marveled at the exquisite Norconian ceiling paintings and started the first art class at the hospital; he utilized a nude female model, which made the sessions very popular. Pictured are fingers coming back to life through the building of a model ship. At one point, Norco had the largest occupational therapy unit in the nation. (LAPL.)

BILLY RAY LANG FROM CHICKASHA, 1944. Physical therapy was another key element of treatment for those suffering from polio, rheumatic fever, and, in the case of this bluejacket, wounds from enemy shrapnel. Naval surgeons at this hospital in particular specialized in shrapnel wound repair. Here, Lang exercises in a "chair designed to renew atrophied leg muscles." (LAPL.)

Loose Lips Sink Ships, 1944. Norco patients line up at a pay phone for a call home or perhaps to score a date. Soldiers did not have cell phones with cameras from which they could send instantaneous commentaries and images of war zones like today. As the sign says, no military information is discussed; to do so at the time of this photograph could get soldiers arrested for treason. (USC.)

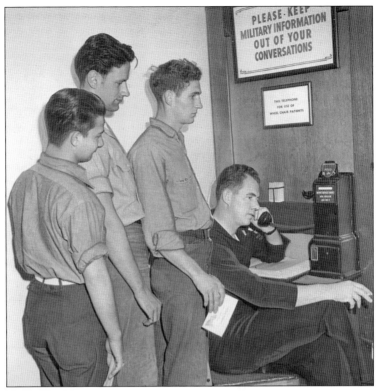

PLEASE·KEEP
MILITARY INFORMATION
OUT OF YOUR
CONVERSATIONS

Delivering Water, 1942. Pictured is the pump house from whence "the hot, healing mineral waters" of the Norconian sprang forth. Needing inexpensive water to operate its growing hospital, the Navy claimed Rex Clark, who controlled all the water in Norco, was charging too much for the commodity, that his delivery system was old and faulty, and as a private company exempt from applicable regulations, could raise rates at any time. (SBM.)

TAKING WATER, 1945. After numerous attempts to negotiate new rates with Clark failed, the Navy condemned and took over the water well fields. Also seized was a swath of land needed to lay pipe to the hospital. New well sites are visible in the left center of this photograph, which shows an area next to the Santa Ana River at the corner of River Road and Bluff Street. (SBM.)

RESERVOIR, 1946. The Navy built or improved at least two known closed reservoirs for its water supply system. The pictured tank overlooks the hospital, is located just north of the facility entrance, and still exists today. Norco residents used the observation tower near the top left of the reservoir as a lookout for any incoming attack on the facility during World War II. (SBM.)

SEWAGE TREATMENT PLANT, 1946. The stress of 5,000 patients, doctors, visitors, and staff took its toll on the old Norconian septic system. The Navy opted once again to seize land owned by Rex Clark to set up what was then a state-of-the-art sewage treatment facility. The red-tile roofing is continued even on this outbuilding far from the hospital. (SBM.)

HEROES, 1944. Citations, commendations, and medals were regularly awarded to patients in the open parade area between the administration building and nurses' quarters. From left to right are Bert E. Zumberge, USMC, (Silver Star); Thomas W. Cooper, USN; Thomas F. Kendrick, USN; Johnnie F. Hamm, USNR; Cpl. Carl F. Eckert Jr., USMC; William F. Lee, USMCR; Robert C. Smith, USN; unidentified; and Captain Jensen.

IN REVIEW, 1944. Nurses march the parade ground in formation for Rear Admiral Edgar Woods, west coast medical facility inspector. In 1942, the hospital had 35 trained nurses; in 1944, the number jumped to 189, and during the peak patient load of 1945, over 200 nurses were on staff. (USC.)

PROVING GROUND, C. 1944. The Spadra Army Hospital near Pomona became Unit IV of the Corona Naval Hospital in Norco and was used as a convalescent facility for patients with a "complete absence of rheumatic fever symptoms" and who needed to prove they could return to active duty. Painting, yard work, building roads, and other chores were the daily routine. Pictured is the base chapel. Spadra was disestablished in 1946. (USC.)

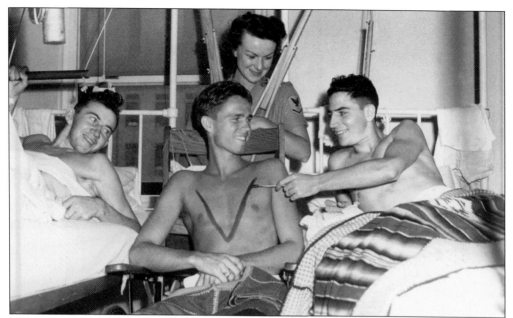

VJ Day, August 14, 1945. On Victory over Japan Day, there was dancing, hugging, and crying in the streets of America and at the Naval Hospital in Norco. According to nurse Beatrice Jankes, "When the Japanese surrendered, I let myself cry for 10 minutes, then went back to work." This photograph, marked "VJ Day," came from a large collection of images marked "USNH Corona," but, sadly, not a single name was listed.

Voyage to Recovery, 1945. As one of the most beautiful naval hospitals in the nation, staged photos, such as this one taken in the so-called ship store, were regularly shot at the Norco facility and used for magazine layouts and posters used to raise support for patients. Additionally, the fascinating short film *Voyage to Recovery* was produced at the hospital. The film illustrated the depth and excellence of care received in naval hospitals.

AT A COST OF $15 MILLION, 1947. Pictured is an aerial south view of the completed naval hospital. Unit I is pictured at lower left center, Unit II in the upper left, and Unit III is in the lower right. While the paint was still drying, a postwar debate to close the facility raged. By 1947, only 511 beds were utilized, and most wards were mothballed. Despite assurances of permanence by Washington, rumors of closing persisted. (SBM.)

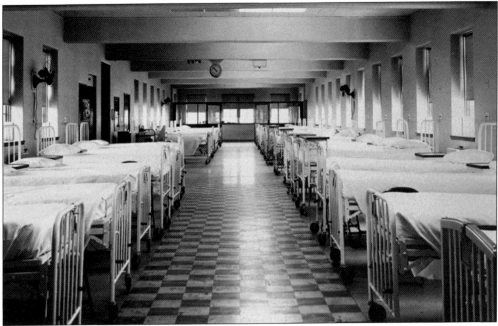

EMPTY BEDS, 1949. By early January, only 150 patients were being treated in Norco. Amidst cries of wasteful spending, federal cutbacks, fears of job loss, and a lack of medical facilities for veterans in the area should the hospital close, a high-profile campaign aimed at filling beds was waged by area civic groups in the hopes of converting the hospital to a veterans' facility.

LAST HURRAH, 1948. In the midst of a nationwide polio epidemic, the excellent therapeutic facilities were made available to afflicted civilians. Pictured is Nedra Eubank working to recover with the help of corpsman William Richards. According to medical officer in command Captain Vann, it would be a waste to close the finest rheumatic fever and polio treatment center in the nation and that "it would never happen." (LAPL.)

CLOSED AND STRIPPED, C. 1948. On November 1, 1949, US Naval Hospital, Corona was disestablished and stripped, with all medical equipment and temporary structures either sold at public auction or taken to other hospitals. Plates, silverware, cooking utensils, towels, and other items dating back to the Norconian were laid on tables, free to patients and staff. The rest, primarily resort furniture, was taken to the base dump, burned, and buried.

DEBATE RAGES, 1950. Three days prior to the outbreak of the Korean War, congressman Richard Nixon spoke against the numerous closures of veterans' hospitals and the resulting lack of medical facilities. Nixon specifically pointed to the expensive completion of the Norco hospital and its wasteful closure within two years. On June 25, 1950, open warfare began in Korea, and Navy plans for Norco were set to change. (USC.)

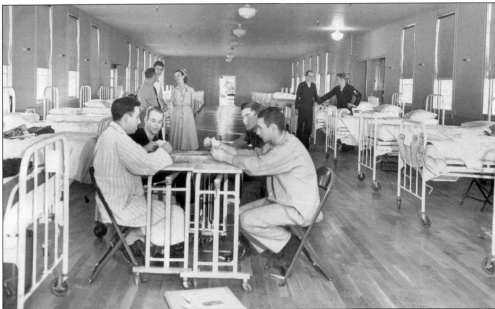

TORPEDOES? 1950. The *Los Angeles Times* announced that the Washington-based National Bureau of Standards would convert the entire hospital into laboratories for the "designing and assembly of a classified Navy tactical guided missile." The long wards, like the one pictured, were thought to be perfect workspaces for torpedo and missile design and development. In anticipation of casualties from Korea, the Navy later decided to reopen the hospital. (USC.)

THEY'RE BACK, 1951. In February, 16 months after stripping the hospital bare, the Navy announced the facility would reopen as a 1,000-bed complex. However, only Units One and Three would be utilized as a hospital, while Unit II would be converted into laboratories for the National Bureau of Standards. Reactivation reportedly cost $2.5 million. Pictured is Florence Rohleder, newly employed at Norco. (Florence Rohleder Christensen.)

INCOMING, 1951. Captain A.B. Chesser (left) took command in July with the first patients arriving from naval hospitals in San Diego and Camp Pendleton in November. The hospital served an area from Los Angeles to San Diego and provided general and clinical medical care to all branches of the armed services. However, Korean War casualties were the primary reason for reactivation. Behind Chesser is the former Norconian clubhouse (USC.)

SPLINTERVILLE REBORN, 1951. Pictured is Unit III, where the first patients were housed in November. This facility consisted of 30 wards with a capacity of 1,040 beds, while Unit I to the east had a capacity of 650 beds. By 1952, roughly 300 officers and enlisted staff and 30 civilian workers were employed at the hospital. Rehabilitation and occupational therapy went into full swing, and a new emphasis in psychiatric care began. (USC.)

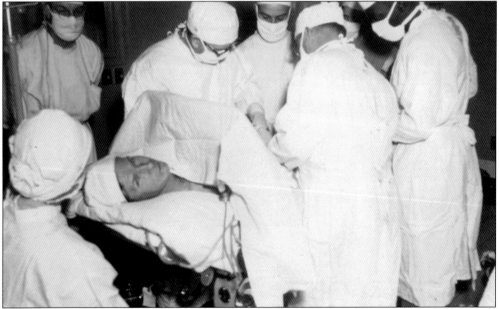

FIRST SURGERY, 1951. American soldiers were well on their way to suffering 33,686 deaths during the Korean conflict, and the hospital in Norco quickly filled. This first operation was to remove shrapnel. As with previous wars, surgical methods improved out of necessity. The primary operating rooms were located on the third floor of the former Norconian Hotel and equipped with state-of-the-art medical equipment. (BUMED.)

Putting On the Feedbag, 1951. In the tradition of the Norconian, famous for exquisite cuisine, the hospital prided itself on feeding its patients and staff well. Seen here is one of three kitchens in use during the Korean War. Each facility was capable of feeding hundreds and was completely self-sufficient. The Navy also built a bakery "topside," from which the smell of fresh bread wafted throughout the complex. (USC.)

Fore! 1955. On the steps of the old Norconian, Capt. Julian Love (left) is presented with new golf clubs for the hospital's patients. In the mid-1940s, nine holes of the old Norconian course were renovated with funding from a series of celebrity golf tournaments. After Love, from left to right, are George Lake (base golf pro), donor Al Santoro, and patients Russell Vohl and Bill Turner. (USC.)

	STATE FILE No			CERTIFICATE OF LIVE BIRTH			REGISTRATION *rural* REGISTRAR'S DISTRICT No. 3300 NUMBER		122
THIS CHILD (TYPE OR PRINT NAME)	1a CHILD'S FIRST NAME MICHAEL		1b MIDDLE NAME IRVINE			1c LAST NAME FRANCIS			
	2 SEX Male	3a THIS BIRTH SINGLE, TWIN, OR TRIPLET Single	3b IF TWIN OR TRIPLET THIS CHILD BORN 1ST, 2ND, 3RD		4a DATE OF BIRTH—MONTH, DAY, YEAR June 15, 1955		4b HOUR 12.17 P.	
PLACE OF BIRTH	5a COUNTY Riverside		5b CITY OR TOWN Corona			X	
	5c FULL NAME OF HOSPITAL OR INSTITUTION U. S. Naval Hospital		5d ADDRESS Corona				
USUAL RESIDENCE OF MOTHER	6a STATE Canada	6b COUNTY Wentworth	6c CITY OR TOWN Hamilton X		6d STREET OR RURAL ADDRESS 6 Dorset Place			
MOTHER OF CHILD	7a MAIDEN NAME OF MOTHER—FIRST NAME Mary	7b MIDDLE NAME		7c LAST NAME Klinko	8 COLOR OR RACE OF MOTHER Cauc.			
	9 AGE OF MOTHER (AT TIME OF THIS BIRTH) 24	10 BIRTHPLACE (STATE OR FOREIGN COUNTRY) Canada		11 MAILING ADDRESS OF MOTHER 1671 Laguna Rd. Tustin, California				
FATHER OF CHILD	12a NAME OF FATHER—FIRST NAME Edward	12b MIDDLE NAME Robert		12c LAST NAME Francis	13 COLOR OR RACE OF FATHER Cauc.			
	14 AGE OF FATHER (AT TIME OF THIS BIRTH) 26	15 BIRTHPLACE (STATE OR FOREIGN COUNTRY) Michigan		16a USUAL OCCUPATION Sgt.	16b KIND OF BUSINESS OR INDUSTRY USMC			
INFORMANT'S CERTIFICATION	I HEREBY CERTIFY THAT THE ABOVE STATED INFORMATION IS TRUE AND CORRECT TO THE BEST OF MY KNOWLEDGE	17a SIGNATURE OF PARENT OR OTHER INFORMANT *Edward R. Francis*			17b DATE SIGNED BY PARENT OR OTHER INFORMANT *June 16 1955*			
ATTENDANT'S CERTIFICATION	I HEREBY CERTIFY THAT I ATTENDED THIS BIRTH AND THAT THE CHILD WAS BORN ALIVE AT THE HOUR, DATE AND PLACE STATED ABOVE R. Derby, M. D.				18a ADDRESS USNH, Corona			
REGISTRAR'S CERTIFICATION	19 DATE RECEIVED BY LOCAL REGISTRAR *June 28 1955*	20 SIGNATURE OF LOCAL REGISTRAR			21 DATE ON WHICH NAME ADDED BY SUPPLEMENTAL NAME RECORD			

THE BABY DOCTOR, 1955. In 1954, Dr. Thomas Lebherz delivered 24 babies in a single 24-hour period in Norco! Famed comedian Groucho Marx was so impressed that he invited the baby doctor onto his television show *You Bet Your Life*. Pictured is the official naval hospital birth certificate for Mike Francis, one of thousands received by happy parents. (Mike Francis.)

LOUIS HAYWARD, 1944. The actor, seen here with wife Ida Lupino, directed *With the Marines at Tarawa*, an Academy Award–winning combat documentary with "gruesome scenes of battle" previously unseen by the American public. For his "meritorious service" Hayward, "wounded, badly shaken, and physically exhausted" from his Tarawa ordeal, received the Bronze Star while recuperating in Norco. President Roosevelt, against the advice of his advisors, approved the film's release so people could understand that war is horrible.

FRATERNIZING WITH AN OFFICER, C. 1955. Frank Day was a corpsman in the psychiatric unit. He fell for Ensign Edna Lorentzen, a US Navy nurse ("starched uniform, pretty woman, and always smelled good"), but as an officer, she was out of his reach. Chemistry overruled military discipline, though, and on each leave, Ensign Edna put Frank in her car trunk and smuggled him off base. The couple is still together 50 years later. (Frank Day.)

NORCO PARADE, 1951. For 70 years, various military and civilian naval base staff members have been active in the community, contributing time and money to all manner of activities, including charities, service groups, and social gatherings. Pictured is a float entry in the 1951 Norco parade; it was named the USS *Naval Hospital Corona*. (Florence Rholeder Christenson.)

CHOPPING BLOCK, 1956. With the war over, politics and the money woes of the federal government once more placed the hospital in jeopardy. The local chamber of commerce and the American Legion lobbied for a full veterans' hospital, but not even a presidential pardon could save the US Naval Hospital, Corona in Norco. In what then-congressman Dalip Saund later said was his bitterest political defeat, the hospital was again disestablished on October 15, 1957.

NO FOLLY HERE, C. 1956. Pictured is one of the last fleet inspections of the hospital. Once reserved for the rich and famous, the Norconian Resort—also known as Rex's Folly—was converted into a preeminent naval hospital where wounded and sick combat vets were treated like kings for almost two decades. According to one longtime caretaker, the place "looks like it was meant for the boys all the time." (Florence Rohleder Christensen.)

Five

MOVIE STARS AND THE BEAUTY BRIGADE

BEAUTY BRIGADE 1944. Wearing their United Service Organizations, Inc. (USO) Camp Show uniforms, these frequent hospital visitors are, from left to right, Linda Darnell, Patty Thomas, Nancy Barnes, and Kim Kimberly. They were part of a large group of actresses who toured with USO troupes, sold war bonds, and visited bases everywhere. The photograph that appears in this book has a history of its own; it was the actual photograph used to announce an upcoming show at the hospital, and a Navy corpsman later carried it throughout the war. (Lesley Mathews.)

KAY FRANCIS, 1943. Although now all but forgotten, stunning Francis (center) was a major movie star in the 1930s. At the outbreak of the war, she began to regularly visit the Corona Naval Hospital, making sure to bring along some of her friends, such as Herbert Marshall (right). With Hollywood so close, it was easy for the biggest stars to visit the boys in Norco. Francis is about to do a radio broadcast to encourage public support for "those boys lying in hospital beds everywhere."

BUNDLES FOR BLUEJACKETS. Wearing Naval Auxiliary Reserve uniforms, Myrna Loy (left) Kay Francis (center), and Virginia Zanuck present Sailors in San Pedro with packages collected in the Bundles for Bluejackets war relief campaign, just as they did in Norco. According to Loy, who was a frequent naval hospital visitor as well as a former regular guest at the Norconian Resort, "We saw untrained kids endlessly stream totally unprepared for war. It broke our hearts." (USC.)

FOR THE BOYS, 1943. Francis, on the old Norconian ballroom stage, was put in charge of morale for the hospital. Towing along other stars, she visited Norco every Thursday, supplying cigarettes, matches, candy, and gum to the patients. Eventually, her operation expanded to include the Long Beach Naval Hospital. Gas rationing curtailed the visits, and tough-to-get free smokes were available to only the sickest patients! Below, John Payne and Francis sign autographs at a grand military ball, which the two stars organized in Hollywood to benefit the naval hospital in Norco. Anyone who was anybody showed up to give autographs and signers included Humphrey Bogart, Cary Grant, James Cagney, Boris Karloff, Clark Gable, Spencer Tracy, Lucille Ball, Katherine Hepburn, and many, many others. However, the real guests of honor were the soldiers themselves.

BATTLING CORSUN, 1942. Kay Francis threw a Bundles for Bluejackets bash and invited several patients from the naval hospital. While dancing with Constance Bennett, Sailor Hyman Corsun (pictured) was tapped by a civilian aircraft worker wishing to cut in. Rather than relinquish the star, Corsun slugged the worker and said, "The Navy never gets anything but coffee and doughnuts while the civilians get all the dances with the movie stars."

BLONDE BOMBER, 1943. A soldier once asked Carole Landis (center), a pin-up girl favorite, to "just sigh" into a microphone. Landis entertained soldiers worldwide and wrote of her experiences in the book *Four Jills in a Jeep.* Pictured in the Hollywood Canteen with two Norco patients, Landis alternated Sundays with Kay Francis at the famed serviceman's club, the Angel Box, visiting with Norco patients. (USC.)

COMMODORE CORYDON M. WASSELL, C. 1945. While a patient in Norco, the man President Roosevelt described as "almost like a Christ-like shepherd devoted to his flock" wrote his story; he made sure 12 wounded Battle of Java Sea survivors got to Australia just as the Japanese were closing in. Gary Cooper starred in the Cecil B. DeMille movie with, at Wassell's request, all proceeds going to the Navy Relief Fund.

JACK BENNY, 1944. The comedian brought his popular radio show to Norco and joked about his own Great Lakes Naval Station days, which he claimed were "almost as bad as being stationed in Corona." The feigned tightwad left behind dozens of gifts for the patients, and when informed that not everyone got a present, the comedian replied, "Get whatever you need, and just send me the bill." (Bison Archives.)

THREE STOOGES, 1940. In the short *You Nazty Spy!* (pictured) Moe Howard was the first American actor to lampoon Hitler. The Stooges were regulars at the hospital, and before a show, they would roam the grounds pretending to be doctors, much to the delight of all. Larry Fine's son would later join the ranks of corpsmen stationed at the hospital. (Academy of Motion Picture Arts and Sciences.)

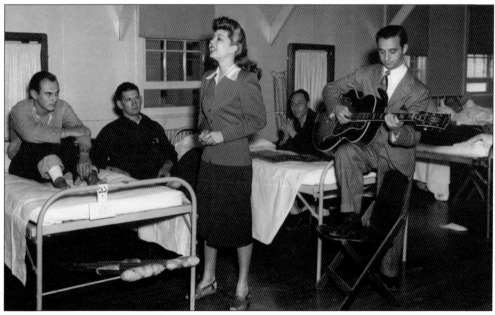

FRANCES LANGFORD, 1945. The popular entertainer, shown at Port of Embarkation Hospital, opened a *Purple Heart* radio show broadcast from Norco with: "Today we're in the place the men call the most beautiful hospital in America. One man told me, if you've got to get it, this is the spot to get over it," and, "[The hospital] is still a luxurious haven for these men who could take it and did."

TIRELESS, 1944. Frances Langford appears in Unit III with (from left to right) Adolph Menjou, Ken Carpenter, and Tony Romano. She sang with the Bob Hope radio show and traveled with the comedian's World War II USO show throughout the world. Langford wrote a weekly column called "Purple Heart Diary" in which she made many a plea for public support of the wounded and sick recovering in hospitals across America.

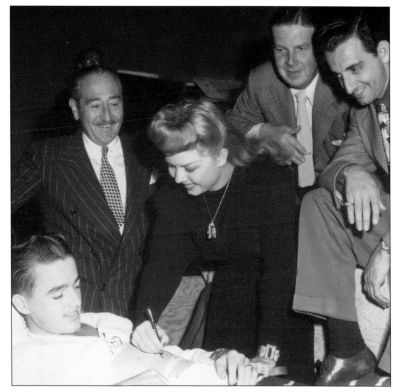

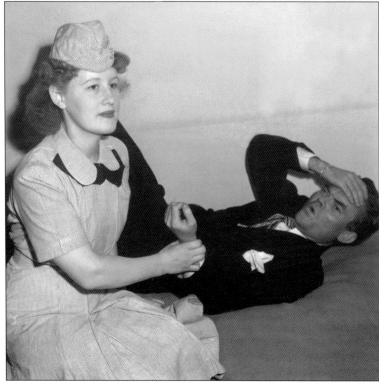

WACKY, 1945. Spike Jones and his City Slickers appeared as guest stars on Frances Langford's Naval Hospital broadcast of her *Purple Heart* radio show. Years before, Jones began his career as the drummer for the old Norconian house band. He is seen here with Navy WAVE Eunice Richardson, who remembered Jones as very "excitable" and "perhaps a little crazy." She also remembered the "boys loved the show." (Eunice Richardson.)

HALL OF FAMER, 1946. One of baseball's most colorful characters and the new owner of the Cleveland Indians, Bill Veeck, is at Norco awaiting the amputation of his infected right foot, the result of an injury received as a Marine at Bougainville. Navy WAVE Vickie Dean said, "My boss was always trying to get me to date Veeck, but he was out of my league."

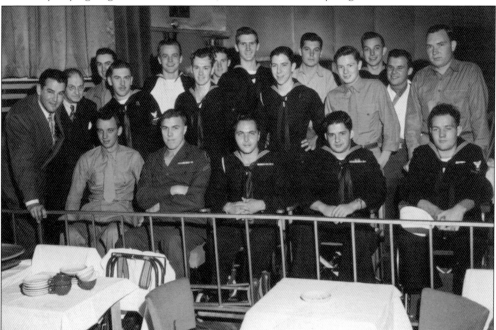

BREAKFAST CLUB, C. 1946. Patients are sitting in on a radio broadcast of Don McNeill's *Breakfast Club* in Hollywood. McNeil (far left) personally saw to it that as many leathernecks and Sailors as possible got a taste of Southern California. Former patient Chuck Roth (second row, third from right) remarkably remembers almost every person in this photograph after more than 60 years. (Charles Roth.)

A Night to Remember, 1944.
Patients were bussed to the
Masquers Club in Hollywood and
greeted by Western star William S.
Hart, the host of a dazzling shindig
in their honor. It is not known who
the lucky fellow was who got these
signatures on this Disney-designed
program, but clearly, the stars
were out that night. Cary Grant's
signature appears to the right.

George Raft, 1942. The tough-guy
actor stated, "Now it's going to be up
to us to send to the men here and
abroad real, living entertainment."
Raft was the first entertainer to
perform for the boys at the new
hospital; over 300 patients crowded
into the old Norconian garage to
watch "Whizzer" Phelps pummel
Joe Ryan for 10 rounds as part of
a so-called sports extravaganza.

NOW THAT'S ENTERTAINMENT, 1945. Naval hospital patients and staff performed many Lee Gangway Nights (a performance of songs sung from the sheltered side of a ship) at Norco, as shown. During World War II, nightly entertainment was offered: Doug Teeple's Music Makers performed often, Mrs. Zane Grey spoke of her late husband's fishing exploits, Johnny Carimi would sing "Lake Norconian Waters," and James Cagney could sometimes be caught hoofing a few steps.

"THE CAMELS ARE COMING, HURRAY, HURRAY!" RJ Reynolds Tobacco Company's road show arrived in Norco on August 20, 1942. Starring in the show were singers Pinky Tomlin and Jeanne Carroll, "versatile musical group the Texas Rangers," and the Paul and Paulette trampoline act. Pictured are Kay Carroll and Gerry Davis, the famed "Camelettes" who closed the show by passing out free cigarettes to patients.

Marlene Dietrich and Kay Kyser, 1942. The zany bandleader famed for his Kollege of Musical Knowledge successfully toured military bases and hospitals across the United States with sultry film star Marlene Dietrich. This photograph was taken at nearby Camp Haan. One week later, Kyser would bring this same show, featuring Linda Darnell, Ginny Simms, Kay Francis, and Carole Landis, to the naval hospital.

The Old Professor, 1942. Bandleader Kay Kyser was hugely popular during World War II and was said to be the first bandleader to have appeared before servicemen. His shows were over the top; Kyser and his band wore wacky costumes and played wonderfully bent instruments. Unit I was under construction at the time of this photograph and is visible in the background. (Florence Marlow.)

BEAUTIFUL DUO, 1942. Jinx Falkenburg (left), a top model and talk show pioneer, and Shirley Patterson, heroine of Western and sci-fi B movies, were tireless USO performers and regular visitors to Norco's naval hospital. Though all but forgotten today, they were hugely popular pin-up models at the time. Thousands of soldiers carried photos of these girls posing in bathing suits and uniforms, sitting on tanks, in planes, and with machine guns. (LAPL.)

BIG BANDS. Pictured is Claude Thornhill's Navy Band, which toured naval hospitals and bases throughout California and overseas. The band featured some of the best musicians in the business. Other big bands to hit Norco were Tommy Dorsey, Will Osborne, Rudy Vallée (and his Coast Guard Band), and Harry James. Several radio programs featuring big-name orchestras originated from the hospital, including Coca-Cola's *Victory Parade of Spotlight Bands*.

BOB HOPE AND BING CROSBY, 1946. Crosby and Hope, pictured at right clowning at Lakeside Golf Course, played several tournaments across Southern California to raise funds for equipment for patients wishing to play on the newly renovated course in Norco. For 50¢, spectators watched the entertainers play 18 comic rounds of golf. Both Crosby and Hope were guests at the old Norconian and later entertained at the naval hospital. In his book *Corpsman, Corpsman!*, corpsman "Shifty" Lucetti remembered a Bob Hope Christmas show that was broadcast from Norco. Hope's first military show was at nearby March Field, and he later broadcast from Camp Anza. Intriguingly, the location of the photograph below has been credited as New York, Hollywood, and Norco. While it is difficult to know where the image was snapped, the people who appear in it are unmistakably identifiable. Hope sits at center with Frances Langford (left), Claudette Colbert (right), and Jerry Colonna (upper left with mustache). (Both AMPAS.)

JASCHA HEIFITZ, 1942. Billed as the "world's most renowned violinist," Heifitz played a special benefit for patients and staff at the hospital. Those planning to attend were forewarned that the doors would be closed at 8:00 p.m., and no one would be admitted "while a number was being played by the great violinist." Norco was definitely a place of class, elegance, and refinement on June 30, 1942.

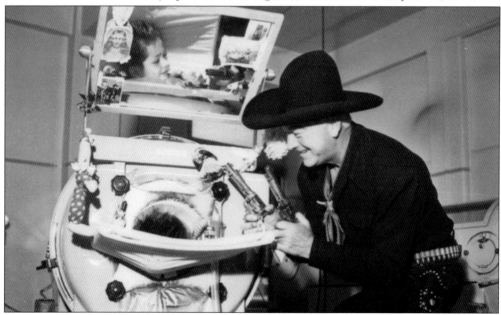

HOPALONG CASSIDY, C. 1947. During the polio epidemic of the late 1940s and early 1950s, William Boyd (who played the television character Hopalong Cassidy) regularly visited children in hospitals all over Southern California, and as seen in this photograph, Norco was no exception. Gene Autry, Roy Rogers, and even Trigger spent time with afflicted children at the naval hospital. (AMPAS.)

Six

War Bonds, Junk, and a Water Buffalo

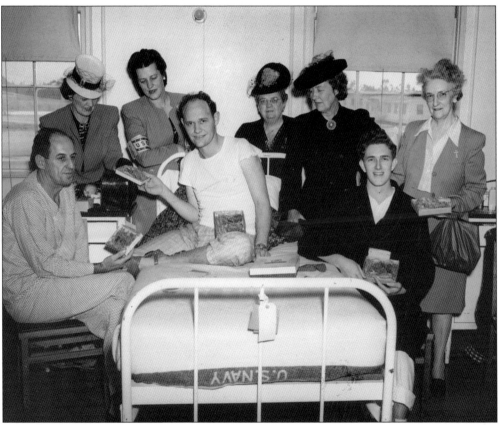

HOME-FRONT ANGELS, 1942. The faces of the home front were, from left to right, Emma Keller, Helen Stanley, Susie Huckins, Evelyn Teague, and ? McMillan. Led by Stanley, these women and others provided patients and servicemen with cakes, lodging, transportation, and dances, and on many occasions, they opened their homes and supper tables. Stanley's daughter Margaret Freeman remembered many occasions when her mother was on the phone seeking a bed for a stranded serviceman. (Corona.)

THE FIRST METAL CASUALTY, c. 1925. Walter Good stands in front of a cannon that sat for many years in the Corona City Park. When World War II broke out, pleas were made for scrap metal of all kinds to build tanks, planes, and bullets. The city fathers voted that this cannon as a symbol would be the first scrap metal to go toward the war effort. (Corona.)

A HERO ON TWO FRONTS, 1942. Standing in the lobby of the former Norconian Hotel is Captain Jensen (left) and Capt. L.B. Marshall. The *Daily Independent* reported that Marshall spoke before the Corona Rotary to discuss, among other things, a "colored serviceman wounded while on duty at Pearl Harbor" who was refused service in a local restaurant. The officer stated that residents must handle this situation. Corona women demanded that service be provided to any and all "fighting boys," and it was. (USC.)

THE GRAY LADIES, 1943. These volunteer saints, organized under the American Red Cross by Helen Stanley (fifth from left) and Mrs. Jensen (center in dress) provided countless hours of non-medical services to patients in Norco. Nettie Whitcomb (sixth from left) ran the Corona Plunge. Upon hearing of the first Hispanic to die in World War II, she refused to continue the policy of segregated swimming hours for Mexican-Americans. (Corona.)

HOMEMADE SWEETS, 1943. Gene Carr of the United States Rubber Company gathered hundreds of pounds of candy for distribution at the three service hospitals near Los Angeles and in particular Norco. Companies of all kinds donated candy, radios, sweaters, and blankets—almost anything a recovering patient might need. In 1945, Corona and Norco organizations claimed to have donated 10,000 cookies! (LAPL.)

BANDAGES FOR THE BOYS, 1942. Women's clubs, the Red Cross, and employees from a wide variety of businesses throughout Southern California joined together on a weekly basis to make surgical dressings for service hospital patients and servicemen oversees. Shown is the Red Cross Unit from J.W. Robinson Department Stores preparing bandages for patients in Norco and Long Beach. (LAPL.)

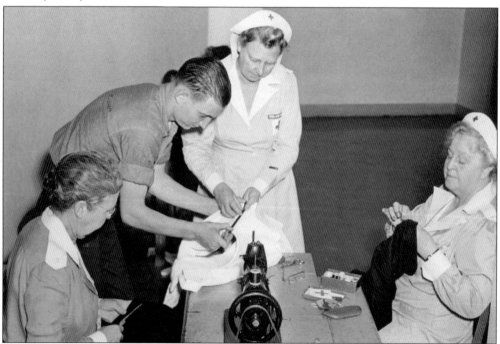

THE DUCHESS, 1943. Margaret Freeman stated that Janet Gould (right) earned a royal moniker because of her custom of "holding court" while serving afternoon tea to visitors at her Victorian home on South Main in Corona. Shown with two other Gray Ladies, Gould is teaching a patient how to sew, likely as part of the hospital's physical therapy program. Gould later became a local historian, writer, and lecturer. (Corona.)

KID POWER, C. 1941. Likely no piece of scrap was left after hundreds of Corona and Norco boys and girls scoured the neighborhoods for auto parts, old hoses, tires, fence posts, wrought iron, gloves, bathing caps, rubber raincoats, and much more; tons of junk were collected. Very likely, most was useless; however, Roosevelt's true goal was to build morale, and it worked. (Corona.)

SCHOOLS AT WAR, C. 1943. The US Office of Education and the US Department of the Treasury organized a competition for students "to work toward victory." Schoolchildren were encouraged to help by purchasing savings bonds and "saving money and materials for the war effort." State winners were chosen based on activity scrapbooks and received a brick from Independence Hall in Philadelphia. Pictured are students from Corona Junior High School, which was attended by residents of Corona and Norco. (Corona.)

SANDS OF IWO JIMA, 1946. The mass production of the amphibious tank—known by the military as a landing vehicle tracked (LVT)—contributed significantly to the war effort. A celebration was held to commemorate the 10,000th LVT manufactured in the United States. During the ceremony, the 10,000th LVT, a Water Buffalo christened with water and sand from Iwo Jima to mark the importance of LVTs in that battle, rolled off the assembly line and into Lake Evans in Riverside, California. Corona Junior High bond sales paid for the tank. Shown from left to right are actor and master of ceremonies Edward Arnold, MGM star Jeanette MacDonald, unidentified, admiral E.L. Cochran (chief of the Bureau of Ships), and unidentified. (Corona.)

GROUND OBSERVATION CORPS, 1943. Corona had five reported observation posts, and Norco had three. The towers were under Army supervision, but constructed by local residents and paid for with scrap metal. Corona was feared to be a perfect target because of circular Grand Boulevard. Until the towers were placed on inactive status after the final drill on May 31, 1944, they were manned by 500 observers 24 hours a day. (Corona.)

A Woody to Remember, 1946. The naval hospital utilized buses to get patients like Charles Roth (second from left) to Corona, Riverside, and Hollywood. However, most fondly remembered were the "Woodies" driven by the Red Cross, Naval Auxiliary Reserve members, and local volunteers. The naval hospital had several of these donated cars as well as access to cars from other Red Cross chapters, like this one from Pomona. (Charles Roth.)

Red Cross Racers, c. 1946. These volunteer drivers not only transported patients but also cared for them. It was not uncommon for a patient passenger to suddenly get sick, fall, or suffer a number of other ailments, and these ladies were there to assist. They also picked up patients from other hospitals and family friends, and along the way, they delivered not a few babies while trying to make it to the hospital. (Charles Roth.)

RAISE THE FLAG, 1948. Charles E. Johnson (carrying flag) marches with the Norco American Legion in the first Norco Fair Parade. He was representative of many Norco and Corona patriots. During the war, Johnson became an air raid warden, sold bonds, visited at the hospital, and transported patients. According to his daughter, "No serviceman on foot ever needed a ride if dad drove by." (Edna Johnson Velthoen.)

CONFUSED STEWARTS, 1943. Corona hero James C. Stewart (center) was nationally confused with that "other fellow," pilot and actor Jimmy Stewart. James C. received the Distinguished Service Cross for leading his nine-plane squadron against 50 enemy planes. Another Corona boy, Harry Sampson received the Silver Star for single-handedly holding a hill against the Nazis. Dozens of local boys received medals and citations for valor. (Corona.)

NAVY DAY, OCTOBER 27, 1943. From 1922 until 1949, Navy Day, while not a national holiday, was a day when Americans honored their Sailors. The Navy League of America suggested October 27 in honor of President Teddy Roosevelt's birthday as the official date. Corona and Norco honored their naval hospital with brass bands and speeches; pretty much every business in town sent newspaper congratulations. (Corona.)

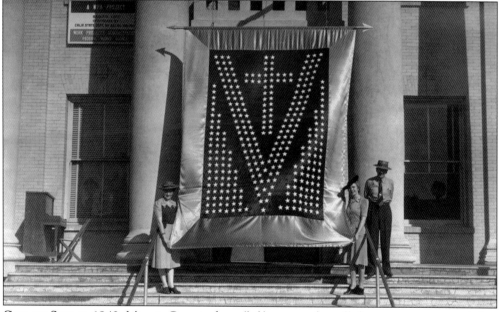

CALL TO SEW, C. 1942. Minnie Cunningham (left) put out the call that women were needed to sew stars representing each Corona or Norco boy in the military onto this Corona Victory Service flag. The goal had been to present the flag in July of 1942; however, "the contacts committee uncovered twice as many names as were originally submitted." By June of 1942, over 300 men in Corona and Norco had enlisted. (Corona.)

ABOVE AND BEYOND, C. 1944. Corona-Norco boys were tough. Second lieutenant Phil Newhouse, pictured, was held for months behind enemy lines before escaping captivity. Brothers Max, Richard, William, John, and Jack Nugent all served in World War II. C.W. Garner was the first Sea Bee on Iwo Jima. Major Fred Payne received the Navy Cross, and corporal Rosario V. Lerma received every award for bravery except the Medal of Honor. (Corona.)

DIXIE'S NAVY PRESS SHOP, C. 1946. Dixie Underhill opened this shop at the corner of Sixth and Hamner Avenue in Norco with the hope of capturing business from the nearby Navy hospital. Her husband, Hugh, was a very popular area constable. This building later became the infamous and wild Red Barn, Granny's diner, and today, Maverick's Steak House. (Debbie Stuckert.)

WAR WOUNDED FUND, 1951. Standing in front of the recreation center, happy Marines and Sailors wave the checks they received from the *Examiner*'s War Wounded Fund. Celebrity extravaganzas raised the thousands of dollars handed out to recovering veterans at hospitals all over California. (USC.)

CORONETTES, C. 1946. Edna Johnson (center, facing camera) is dancing with a Sailor in the old Norconian ballroom during one of the many dances attended by the Coronettes, a girls' social club. Supervised by Helen Stanley, the original group of 50 girls disbanded in 1947 as the hospital downsized—and 48 of the girls had gotten married! (Edna Velthoen Johnson.)

JOE DOMINGUEZ POST 742, 1948. Upon returning home, Hispanic veterans were encouraged by some members of the all-white Corona Post 216 to form their own organization; they did exactly that and named it after Jose Dominguez, who was killed in the Pacific on December 27, 1943; he was 22 years old. Onias Acevedo says that their early years as a post were spent "burying our boys, not worrying who hated us." (Corona.)

GOODBYE, 1947. As the hospital downsized, the number of servicemen to hold dances for had dwindled. So, a Farewell to the USO formal dance to the music of Ted Ward and his orchestra was held. Corona USO founder Helen Stanley (fourth from right in the corsage) smiles as an era ends. For years after, this marvelous lady received grateful notes from veterans who she had served so well. (Corona.)

Seven

MAN THE TORPEDOES

COLD WAR WARRIORS, 1951. Developing everything from radio-control bombs to pilotless aircraft in secret laboratories in Washington, DC, the National Bureau of Standards (NBS) was cramped into tiny Quonset huts when it "officially designated the Missile Development Division." Workspace was desperately needed, and when the Norco hospital was "declared surplus," the long hospital wards, such as those shown from Unit III, were deemed ideal for "perfecting missiles." Congress eagerly funded the move.

SICK BAY TO BOMB BAY, C. 1953. When the decision to re-establish the hospital was made, the National Bureau of Standards announced it would still convert the surplus tubercular wards into Missile Development Division laboratories. However, remodeling cost overruns forced the jet engine laboratory, the wind tunnel, and the associated staff to remain in Washington. Above, a new tower, "used to point radar systems out the windows for an unobstructed view of targets," rises behind the former staff quarters. Below, the glass between the weeping mortar porch pillars was replaced with windows and very "temporary" infill walls. The interiors were redesigned with non-bearing walls to create laboratories, machine shops, computer rooms, and other workspaces. Nervous residents were assured by spokespersons that "we shan't be firing any missiles." (Corona.)

CREAM OF THE CROP, 1951. The Missile Development Division move west was a huge event with 250 scientists, engineers, and physicists arriving in Norco. Local talent was also used; pictured is nearby Eastvale resident Loren Meissner, who began renovating the base golf course while still in high school, moved up to supervisor at the Naval Ordnance Lab Corona Computer Center, and eventually became a professor at the University of San Francisco. (Loren Meissner.)

BAT, 1952. Because noisy World War II aircraft could not sneak up on German submarines, the NBS created the Navy BAT (pictured), the world's first operational air-to-surface missile to use radar for guidance. The Bureau of Ordnance history states the World War II BAT "ranked with the atom bomb and the proximity fuze as one of the few entirely new weapons in World War II." (Jerry Stewart.)

PRECISION MEASUREMENT, C. 1952. According to a statement from Washington: "The most important activity at the Corona Laboratories will be the development of guided missiles. Every phase of missile development will be covered—from theoretical and applied research to construction of experimental parts and units." Pictured is a technician "measuring attenuation of a shielded enclosure" to determine how the vibration of a missile being fired will affect equipment behind safety barriers. (Jerry Stewart.)

NAVAL ORDNANCE LABORATORY CORONA (NOLC), 1953. The secretary of defense announced that weapons research and development was more a function of the military than of the NBS. The lab was transferred to the Department of the Navy, and the base underwent the first of many name changes. Oddly, the Navy continued to use Corona as part of the name despite the base location in Norco. (NSWACSBDC.)

TERRIER MISSILE, C. 1955. Launched from the USS *Mississippi*, the Terrier (above) was the Navy's first surface-to-air guided missile. The crucial responsibility of accurately and objectively evaluating full-scale shipboard missile firings was assigned to scientists in Norco. In 1955, Dr. Stanley Atchison became the NOLC's technical director, and Aaron Powers headed the Missile Evaluation Department (MED). Pioneering advances were made in the development of guidance and telemetry systems, computer components, infrared spectroscopy, and other areas. The MED eventually became responsible for evaluating each new Navy missile as it was introduced. Pictured below, from left to right, are Aaron Powers, Dr. Stanley Atchison, Jerry Stewart (Computer Division director), and Captain F.J. Heiler. Powers would soon become the technical director of the Fleet Missile Systems Analysis and Evaluation Group (FMSAEG), one of the most crucial divisions in the Navy. (Jerry Stewart.)

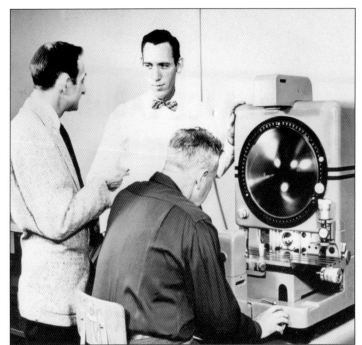

NO ROOM FOR ERRORS, c. 1954. MED scientists in Norco were tasked with appraising weapons manufacturers, evaluating their quality of production and ability to fulfill contracts in accordance with specific requirements and inspection procedures. The Missile Evaluation Department also monitored and assessed manufacturer test firings, inspection procedures, calibrations programs, and so on. Here, an optical comparator used to inspect manufactured parts is being adjusted. (Jerry Stewart.)

DOES IT KEEP ON TICKING? This photograph from around 1957 shows how Norco scientists were tasked with making sure weapons worked when needed in combat. Special laboratories were created to simulate extreme conditions comparable to those found in any place where the Navy might engage in battle. Missile components, computer parts, switches, lenses, and hundreds of other items were heated, frozen, dumped in salt water, dropped, and hit with hammers. (NSWACSBDC.)

FUSES, C. 1964. The technology to produce a direct hit with a torpedo during World War II was nonexistent, and even today it is a challenge. The NOLC Fuze Department researched and developed technologies to achieve detonation close to the target, which created more damage over a wider area. Pictured is an air gun used to test the effects of velocity on fuze components and detonation. (NSWACSBDC.)

A PILE OF PAPER, C. 1969. The MED processed huge amounts of data on production quality, missile histories, field tests, and so on. To process, store, and retrieve the unprecedented amount of information collected, MED pioneered the use of large-scale computers in development in the 1950s. The succession of computers used was state-of-the-art and reflected "the progress in computer technology over that period." (NSWACSBDC.)

DESERT MISSILES, C. 1960S. With near-perfect flying weather year round and almost unlimited visibility, China Lake was the Navy's primary land-based missile test range. Proximity to that isolated base was a major reason for the placement of the Missile Development Division in Norco. Pictured above at China Lake is a NOLC flyover test, part of a guided missile fuze development project. Below is likely a view of a missile proximity fuze test to assess detonation reliability. A surplus military rocket would be mounted with fuze components and propelled at high speeds on a fabricated sled along the Supersonic Naval Ordnance Research Track (SNORT). Speeds of 3,000 miles per hour or more could be reached. At times, a missile-tipped rocket was fired at a 10-foot-thick, rebar-reinforced concrete block wall placed at the end of the four-mile track—*ka-boom*! (Corona.)

MUSIC AND MISSILES, C. 1964. At China Lake, Norco physicist Jack Henderson (center) examines what is left of a BAT after hitting the ground at 200 miles per hour. Built by Wurlitzer, which was best known for manufacturing musical instruments, the BAT was a plywood glide missile with no power. BAT bodies were used for over a decade to carry all types of experimental guidance systems. (Jack Henderson.)

CREATIVE GENIUS, C. 1960S. By all accounts, Navy scientist Jack Richer was very smart and wonderfully screwy. He loved to make clocks go backwards and had a fondness for the old electric Red Cars that once ran on tracks all over California. Here, Jack reclines on rails he installed near his home in Norco. Many a child (including the author) took a ride on his personal Red Car. (Bill Erickson.)

ELECTRONIC BRAINS, 1953. At the 1953 Western Computer Conference, NBS Corona technical director Dr. R.D. Huntoon stated, "In the computer art, we ain't seen nothing yet!" In 1949, the National Bureau of Standards produced Standards Eastern Automatic Computer (SEAC), the fastest general-purpose, electronic computer in existence. The so-called computer explosion that began in Washington would become the key to guided missile research and development. By 1953, Norco was utilizing the Goodyear Electronic Differential Analyzer (GEDA) pictured above. That same year, the National Advisory Committee for Aerodynamics (eventually NASA), utilized the GEDA to support what was likely the first high-speed flight simulator for research purposes. Pictured below is one of the Iconolog cinetheodolite film readers used by Norco scientists. Film from multiple cinetheodolite cameras was combined and read to determine test missile trajectory, accurate to within approximately one meter. (Jerry Stewart.)

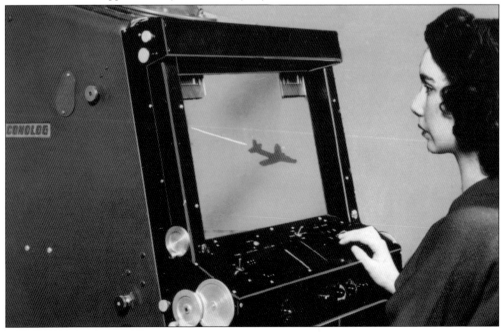

COMPUTING POWER. A state-of-the-art Reeves Electronic Analog Computer (REAC) was initially installed in Norco. Dr. Loren Meissner remembers, "The first project I worked on (1951) was adding guidance to air-launched antisubmarine torpedoes. Digital computers were just becoming available, and the Navy wanted someone from our lab to explore the advantages (if any) of doing these simulations digitally. The National Bureau of Standards was doing forefront research, looking for the best ways to use the new analog and digital computer technology for the nation's needs, and in the early postwar period some important applications included flight simulations such as those being undertaken at the Norco lab." Above is a NOLC-developed Teledac decommutator to handle data from the Polaris missile. Below is the Universal Automatic Computer (UNIVAC) 1180; the computer correctly predicted the 1952 presidential election, much to the dismay of the Democrats. (Corona.)

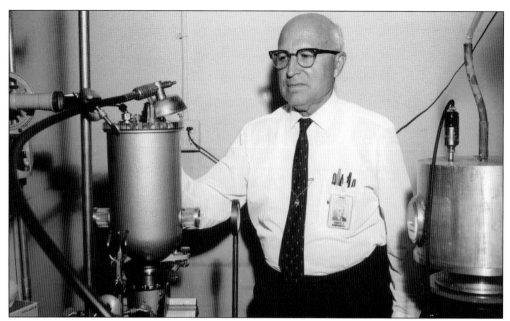

Dr. Curtis Humphreys, 1967. Perhaps the most renowned scientist to move to the West Coast in 1951, Dr. Humphreys was known worldwide for his work in atomic spectroscopy. He properly placed uranium on the periodic series and first measured the sixth series of atomic hydrogen (known as the Humphreys Series). Dr. Humphreys and his staff also pioneered magnetic tape for use in digital computer storage devices, among dozens of other innovations and inventions. (Corona.)

Patents, 1967. Warren R. Hoelzen receives a check for his share of three patents from Naval Weapons Center Corona captain E.B. Jarman. Inventors were allowed to have patents listed in their name, but the federal government retained use under a royalty-free license. Perhaps the most prolific inventor at Norco was Frederick C. Alpers, who received well over 50 patents. (Corona.)

Scouting, c. 1953. Many naval officers and staff from both the hospital and laboratories had children in various scouting organizations. For over 50 years, the Navy Sea Cadet Division has been active in Norco, and thousands of Boy Scouts, Girl Scouts, Campfire Girls, and other youth groups have camped around Lake Norconian. Here, Cub Scouts visit the NOLC exhibit at the Los Angeles County Fair. (Corona.)

FMSAEG, 1963. Admiral Eli Reich, commander of the Surface Missile Project, created the Fleet Missile Systems Analysis and Evaluation Group (FMSAEG). Simply put, torpedoes and missiles were either firing too soon, too late, or not at all, and Reich needed a group of completely independent analysts to find out why—thus the Missile Development Department became FMSAEG. Cdr. G. Estes (far left) and Adm. Eli Reich (second from right) appear in the photograph. (NSWACSBDC.)

EXPANSION, C. 1964. With the closure of the hospital in 1957, the laboratories began to expand into the old Norconian garage, and the former chauffeurs', WAVES, and corpsmen's quarters. Standing in front of the new FMSAEG Department Offices (former corpsman quarters) are Cdr. Glenn Estes (holding flag, left) and Aaron Powers, (holding flag, right). For 30 years, this division was entrusted with assessing and evaluating Polaris missile performance. (NSWACSBDC.)

MISSILES, 1976. Joe Terry sits astride a Talos missile exhibited during one of the annual Armed Forces Day open houses on the naval base. Norco scientists contributed significant technical direction, research, and development to the missile program. They pioneered many guidance system and fuze innovations, conducted quality and reliability assessments, and much more. Missile systems they contributed to included Talos, Terrier, Tartar, Standard Arm, Aegis, Sparrow, Sidewinder, Phoenix, Polaris, Poseidon, Trident, Tomahawk, Bullrup, Walleye, Shrike, and others. (NSWACSBDC.)

ARMED FORCES DAY, 1965. A popular, semi-annual tradition has been for the Navy base to "suspend the secret phases of its work" and welcome the general public in for a look at the facilities and some of the non-classified projects. Here, a US Marine Corps helicopter lands on the east side of the laboratory campus. (Bill Erickson.)

DRONE TEST, 1965. Here, during the Armed Forces Day open house, visitors could see how Norco scientists measured the radar return from a drone with the sphere, which provided calibration. The goal was to measure how detectable an object was using radar (known as radar cross section). Through the years, Navy scientists have displayed some pretty amazing exhibits on Armed Forces Day, including tanks, missiles, jets, electromagnetic demonstrations, and much more. (Bill Erickson.)

BLAST OFF, C. 1965. This NOLC testing tower, built overlooking the laboratories, was visible throughout Norco; local children used to imagine that on any given morning, this rocket ship was going to the moon. However, it was no match for the famous Norco Santa Ana winds. On March 23, 1966, as the tower rocked back and forth, lab employees placed bets on what time it would fall. (Navy.)

CAME TUMBLING DOWN, 1965. The *Daily Independent* wrote, "In theory, the $67,000 foam testing tower can withstand gale winds up to 110 miles per hour . . . But the top half toppled yesterday, after hours of teetering back and forth." Commanding officer Captain Jarman commended Bill Erickson, John Everly, Earle Johnson, and Barry Todd for videotaping the expensive collapse. (Corona.)

NAVY FIRE DEPARTMENT, 1969. There were two fire stations during the years when the naval laboratories and hospital coexisted, and Norco depended on Navy firefighters on many occasions to put out local fires. Pictured is a children's field trip to the base with jaunty NOLC-FMSAEG fire chief J.R. Rabourn to the left. The first Norco Volunteer Fire Department truck was donated by the Navy hospital as it was closing. (Corona.)

FUZE LABORATORY, C. 1965. Between 1956 and 1965, several laboratory and storage buildings and test structures were constructed on the grounds, including a microwave systems field test facility, an antenna test facility, and this building, used to develop, test, and assemble fuzing systems. Plans were also made to build a flight simulator test facility inside this top-secret lab, but whether that occurred or not is confidential. (NSWACSBDC.)

WORLDWIDE, C. 1971. Lab scientists, particularly during the FMSAEG era, manned telemetry stations around the world as well as Navy Liaison Offices from San Diego to Washington. For shipboard operations, a mobile laboratory was utilized. Here, from left to right, Bill Dissette, Bill Erickson, and Mark Pearson have just finished a test flight in Puerto Rico aboard the Navy's last World War II antisubmarine aircraft, the *Truculent Turtle*. (Bill Erickson.)

HILL B, C. 1960. The two-story block building was first constructed as a missile component test facility and radio frequency testing unit. Later, it was utilized as part of the Weapons Impact Scoring System (a system used to train every military pilot in America) research and development project. Pictured is a beautiful eastward view of Lake Norconian, the naval laboratories, and in the distance, Norco. (Jerry Stewart.)

SOLAR FLARES, C. 1968. "Norco was the only Observation Point in the United States that could receive afternoon solar radio emissions. Prior to satellite communication networks, solar flares could shut down long distance communications for extended periods of time." Seen here in the 1960s are state-of-the-art solar optical and radio telescopes in front of the former TB ward Corpsman Quarters. (Bill Erickson.)

EMPTY WALKWAYS, 1971. On several occasions during times of federal budget cutbacks and base consolidation efforts, the laboratories were faced with closure. On June 30, 1971, the flag was actually ceremonially lowered, and the base was "for the rabbits now." However, the laboratories did not close as public outcry held sway. This marvelous weeping mortar pillar walkway protects hundred of scientists from the elements to this day. (Corona.)

CHANGE OF COMMAND, 1966. In front of FAMSAEG headquarters (former WAVES quarters), command is passed from Captain Estes to Captain Heiler. Since 1941, there have been nearly 50 different commanding officers overseeing the various base missions. During the hospital years, thousands of enlisted personnel and officers manned the base. Since the 1957 hospital closing, generally less than five military personnel are employed at the facility with primarily civilian contractors doing the work. (NSWACSBDC.)

WHAT'S IN A NAME? Over the years many photographs were taken in front of this beautiful Claud Beelman–designed command headquarters. Pictured in 1963 is a historic group shot taken during the conference that established FMSAEG as the Navy's independent missile assessment agent. Lab name changes over the archway have been constant and include the National Bureau of Standards, the Naval Ordnance Laboratory, the Naval Weapons Center Corona Laboratories, and several others. (Jerry Stewart.)

Eight

STILL STANDING

THE FAT LADY SINGS, 1961. Standing before the former Norconian clubhouse are Father Galinda, Capt. H.P. McNally, monsignor Matthew Thompson, and another unidentified priest. The pictured Norconian clubhouse (used as the hospital administration building) and several structures had been closed for almost four years, but the battle to bring a veterans' hospital to Norco was still raging. Finally, in November 1961, the hospital was officially disestablished and put up for surplus. (Corona.)

RECOVERY, 1962. Units One and Three were transferred to the state, and the California Rehabilitation Center (CRC), the first state-funded drug rehabilitation program in the United States, was born. Male and female drug offenders could now choose between going to prison or going to Norco for treatment. Guards were known as counselors and prisoners were not incarcerated, but enrolled. Pictured is then-governor Edmund G. "Pat" Brown and pioneer warden Roland Wood. (Corona.)

SANCTUARY, C. 1964. The Navy influence in Norco has had another little-known positive effect. St. Luke's Chapel has literally been in constant use since the day of its first US Navy Hospital Christmas Eve service. Every religion is now served in this simple but beautiful chapel, and it is the only safe haven from gang conflict within CRC fences. (USC.)

REHABILITATION, C. 1970s. Performing in the former Norconian ballroom are professional dancers, who work to get women off drugs through artistic endeavors. CRC pioneered drug rehabilitation through the arts, and some pretty big names performed for prisoners and conducted workshops in theater, music, and dance over the years. Stars included Ruby Dee, Arnold Schwarzenegger, Ossie Davis, and most recently, Tim Robbins and the Actors Gang. (CRC.)

RIOT! 1968. A guard was overpowered, and male prisoners proceeded to rip down and burn the Navy-built dorm rooms, vocational building, firehouse, and guardhouse. Prisoners fought firefighters as they attempted to quash the fire, resulting in extensive damage to former naval hospital Unit III. Norco residents were up in arms over the incident, and prison officials issued assurances that inmate controls would become tougher. (Corona.)

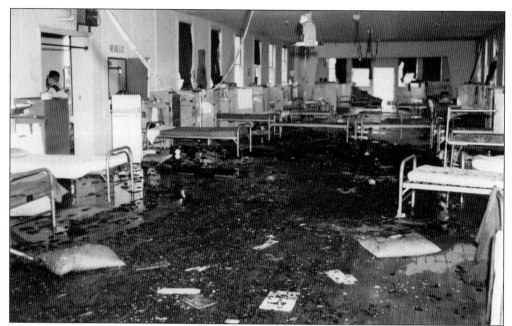

HOSPITAL TO HAVOC, 1968. Shown is one of the wards after the riot. Many within the California prison system deemed the great experiment in drug rehabilitation a failure. Today, the facility—essentially still a hospital campus with wire—has moved away from its original drug rehabilitation focus and now incarcerates a wide spectrum of inmates. Still, there are those who fight daily to rehabilitate the people society wishes to forget. (Gary Judkins.)

FUTURE NORCO COLLEGE, 1985. The Riverside Community College District quickly claimed the 141 acres on the southern end of the Navy base when they were declared surplus. Looking west, remnants of the Crest Drive-In (a Norco landmark), the former Navy landfill, and the Norconian golf course are visible. On this site today are the thriving Norco College and John F. Kennedy High School. (Norco College.)

GOODBYE, C. 1991. In the mid-1990s, the laboratories required more modern and updated facilities. Sadly, historic structures—including the pictured former Norconian laundry constructed by famed architect G. Stanley Wilson—were torn down to make room. Other victims to the wrecking ball were the historic WAVES quarters and the Unit II nurses' quarters. Nevertheless, most buildings from both the Norconian and the naval hospital eras are incredibly intact. (NSWACSBDC.)

HOTEL CALIFORNIA, C. 2005. Legend has it that the CRC, pictured, was the inspiration for the Eagles' hit song "Hotel California." However, the album cover photograph shows the Beverly Hills Hotel, and the band states officially that inspiration came from elsewhere. Pictured in this southwest view are the abandoned clubhouse (upper left), the now-abandoned corpsman quarters (lower left), and prisoner-filled Unit I. (Brigitte Jouxtel.)

GHOSTS, 2005. In 2002, due to "seismic considerations," the old Norconian clubhouse was fenced off and today is empty and disintegrating. A longtime naval security officer claims that on many nights, "a group of well-dressed, 1920s-era people regularly walks from the old hotel [pictured behind barbed wire] through the fences to the lake pavilion where lights get flicked on, music plays, but upon inspection, the joint is empty!" (Brigitte Jouxtel.)

ONE DOLLAR, 2000. After intense lobbying on the part of the city of Norco and California assemblyman Rod Pecheco, and despite strong opposition from the California Department of Corrections, legislation was passed in 2003 that would enable the city of Norco to take the now abandoned Norconian and surrounding grounds for a single dollar. This is a northeast view showing the beautiful 75-year-old landscaping. (City of Norco.)

SURROUNDED, 2003. Despite historic preservation protections and public outcry, the state bulldozed the manicured grounds around the clubhouse and constructed temporary office buildings. According to state officials, Norco "possessed" the clubhouse and the land it directly sat on; however, the area surrounding the building was prison territory and could not be crossed. Barbed wire went up around the majestic and dying Norconian, and any use was rendered impossible as of this writing. (CRC.)

STILL IN NORCO, 2008. State-of-the-art, technologically advanced laboratories have been constructed, and the base remains crucial to national security. Pictured at a recent Pearl Harbor commemoration event are Admiral Herring and Captain Kurtz. Prior to retirement, they were in charge of what is now known as the Naval Surface Warfare Center, Seal Beach, Detachment, Corona—but we know they are standing in Norco. (Brigitte Jouxtel.)

www.arcadiapublishing.com

Discover books about the town where you grew up, the cities where your friends and families live, the town where your parents met, or even that retirement spot you've been dreaming about. Our Web site provides history lovers with exclusive deals, advanced notification about new titles, e-mail alerts of author events, and much more.

MADE IN THE USA

Arcadia Publishing, the leading local history publisher in the United States, is committed to making history accessible and meaningful through publishing books that celebrate and preserve the heritage of America's people and places. Consistent with our mission to preserve history on a local level, this book was printed in South Carolina on American-made paper and manufactured entirely in the United States.

This book carries the accredited Forest Stewardship Council (FSC) label and is printed on 100 percent FSC-certified paper. Products carrying the FSC label are independently certified to assure consumers that they come from forests that are managed to meet the social, economic, and ecological needs of present and future generations.

FSC

Mixed Sources
Product group from well-managed forests and other controlled sources

Cert no. SW-COC-001530
www.fsc.org
© 1996 Forest Stewardship Council

Find Your Place in History.